Buckle Down™

to the Common Core State Standards

Mathematics

Grade 6

This book belongs to: ____

ISBN 978-0-7836-7988-4

1CCUS06MM01 12

Cover Image: Colorful marbles. © Corbis/Photolibrary

Triumph Learning® 136 Madison Avenue, 7th Floor, New York, NY 10016

Printed in the United States of America.

Frequently Asked Questions about the Common Core State Standards

What are the Common Core State Standards?

The Common Core State Standards for mathematics and English language arts, grades K–12, are a set of shared goals and expectations for the knowledge and skills that will help students succeed. They allow students to understand what is expected of them and to become progressively more proficient in understanding and using mathematics and English language arts. Teachers will be better equipped to know exactly what they must do to help students learn and to establish individualized benchmarks for them.

Will the Common Core State Standards tell teachers how and what to teach?

No. Because the best understanding of what works in the classroom comes from teachers, these standards will establish *what* students need to learn, but they will not dictate *how* teachers should teach. Instead, schools and teachers will decide how best to help students reach the standards.

What will the Common Core State Standards mean for students?

The standards will provide a clear, consistent understanding of what is expected of student learning across the country. Common standards will not prevent different levels of achievement among students, but they will ensure more consistent exposure to materials and learning experiences through curriculum, instruction, teacher preparation, and other supports for student learning. These standards will help give students the knowledge and skills they need to succeed in college and careers.

Do the Common Core State Standards focus on skills and content knowledge?

Yes. The Common Core State Standards recognize that both content and skills are important. They require rigorous content and application of knowledge through higher-order thinking skills. The English language arts standards require certain critical content for all students, including classic myths and stories from around the world, America's founding documents, foundational American literature, and Shakespeare. The remaining crucial decisions about content are left to state and local determination. In addition to content coverage, the Common Core State Standards require that students systematically acquire knowledge of literature and other disciplines through reading, writing, speaking, and listening.

In mathematics, the Common Core State Standards lay a solid foundation in whole numbers, addition, subtraction, multiplication, division, fractions, and decimals. Together, these elements support a student's ability to learn and apply more demanding math concepts and procedures.

The Common Core State Standards require that students develop a depth of understanding and ability to apply English language arts and mathematics to novel situations, as college students and employees regularly do.

Will common assessments be developed?

It will be up to the states: some states plan to come together voluntarily to develop a common assessment system. A state-led consortium on assessment would be grounded in the following principles: allowing for comparison across students, schools, districts, states and nations; creating economies of scale; providing information and supporting more effective teaching and learning; and preparing students for college and careers.

TABLE OF CONTENTS

		Page	Common Core State Standards
Introduction		7	
	Test-Taking Tips	8	
Unit 1: The Number System		9	
Lesson 1	Integers	10	6.NS.5, 6.NS.6.a
Lesson 2	Rational Numbers	14	6.NS.6.c
Lesson 3	Absolute Value	18	6.NS.7.c, 6.NS.7.d
Lesson 4	Comparing and Ordering Rational Numbers	22	6.NS.7.a, 6.NS.7.b
Lesson 5	The Coordinate Plane	28	6.NS.6.b, 6.NS.6.c
Lesson 6	Dividing Whole Numbers	35	6.NS.2
Lesson 7	Common Factors and Multiples	38	6.NS.4
Lesson 8	Adding and Subtracting Decimals	42	6.NS.3
Lesson 9	Multiplying and Dividing Decimals	46	6.NS.3
Lesson 10	Dividing Fractions	52	6.NS.1
Unit 1 Practice Test		57	
Unit 2: Ratios and Proportional Relationships		65	
Lesson 11	Ratios	66	6.RP.1
Lesson 12	Percents	70	6.RP.3.c
Lesson 13	Rates and Unit Rates	76	6.RP.2, 6.RP.3.b
Lesson 14	Solving Problems Using Graphs	80	6.RP.3.a, 6.EE.9
Lesson 15	Converting Measurements	85	6.RP.3.d
Unit 2 Practice Test		91	
Unit 3: Expressions and Equations		99	
Lesson 16	Writing Expressions	100	6.EE.1, 6.EE.2.a, 6.EE.2.b, 6.EE.6
Lesson 17	Evaluating Expressions	105	6.EE.1, 6.EE.2.c
Lesson 18	Equivalent Expressions	110	6.NS.4, 6.EE.3, 6.EE.4

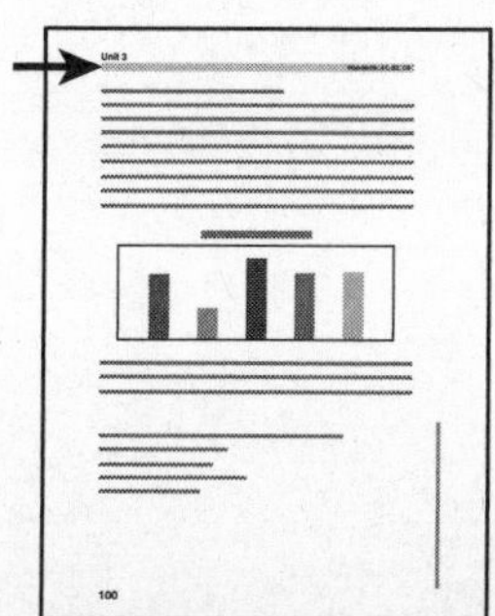

To the Teacher:

Standards Name numbers are listed for each lesson in the table of contents. The numbers in the shaded gray bar that runs across the tops of the pages in the workbook indicate the Standards Name for a given page (see example to the left).

Introduction

Math is all around you. Take this book as an example. Math was used to calculate the margins and the number of pages needed. Math was used to decide where to put tables and other pieces of art. Math was even used to design the cover. Take a look around your classroom—the walls, the pencil sharpener, the windows, and even the plants you see relate in some way to math. You probably use math all the time without even knowing it.

This book will help you practice the math skills that you need in your everyday life, as well as in school. As with anything else, the more you practice these skills, the better you will get at using them.

Test-Taking Tips

Here are a few tips that will help you on test day.

TIP 1: Take it easy.

Stay relaxed and confident. Because you've practiced these problems, you will be ready to do your best on almost any math test. Take a few slow, deep breaths before you begin the test.

TIP 2: Have the supplies you need.

For most math tests, you will need two sharp pencils and an eraser. Your teacher will tell you whether you need anything else.

TIP 3: Read the questions more than once.

Every question is different. Some questions are more difficult than others. If you need to, read a question more than once. This will help you make a plan for solving the question.

TIP 4: Learn to "plug in" answers to multiple-choice items.

When do you "plug in"? You should "plug in" whenever your answer is different from all of the answer choices or you can't come up with an answer. Plug each answer choice into the problem and find the one that makes sense. (You can also think of this as "working backward.")

TIP 5: Answer open-ended items completely.

When answering short-response and extended-response items, show all your work to receive as many points as possible. Write neatly enough so that your calculations will be easy to follow. Make sure your answer is clearly marked.

TIP 6: Use all the test time.

Work on the test until you are told to stop. If you finish early, go back through the test and double-check your answers. You just might increase your score on the test by finding and fixing any errors you might have made.

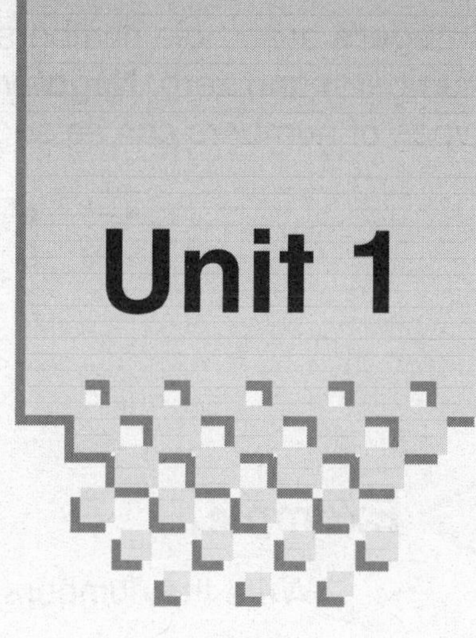

The Number System

You will perform operations with numbers every day of your life. You might need to add to find the total cost of two items. You might need to subtract to find the change owed from a purchase. You could multiply or divide to combine or split groups of items. Not all of the numbers you work with will be whole numbers, however. Whether in the grocery store or on the stock market, numbers will often appear as fractions or decimals. Some numbers can even be negative, like a temperature of −4°C.

In this unit, you will use a number line to show the values of different rational numbers. You will learn how the number line can help compare and order numbers. You will plot and find points on the coordinate plane. You will learn how to find the factors and multiples of whole numbers, including the least common multiple and the greatest common factor of two numbers. You will add, subtract, multiply, and divide decimals. Finally, you will learn about dividing fractions, including using area models to help.

In This Unit

Lesson 1: Integers

Integers are whole numbers and their opposites (positive numbers and negative numbers) and zero. **Negative numbers** are the numbers less than zero. All of these types of numbers can be shown on a number line.

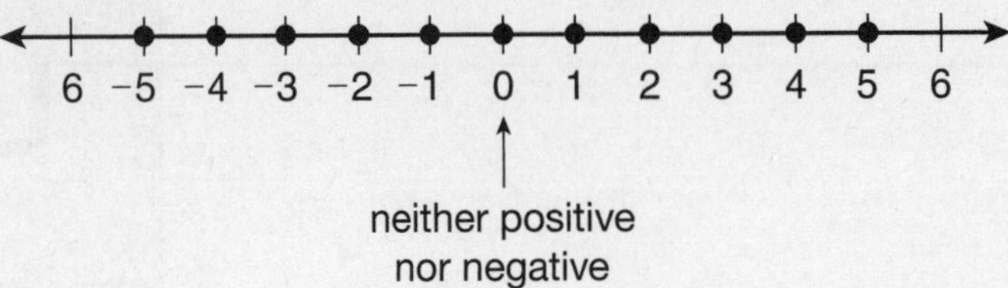

Example

Write the numbers 2, 5, −4, and −1 in their correct place on the number line.

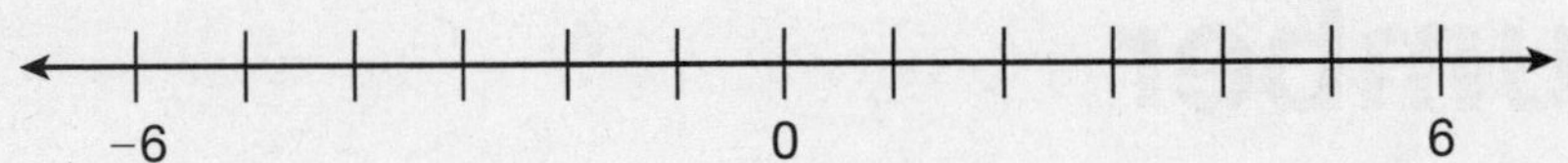

Determine what each mark on the number line represents. On the number line above, each mark represents 1 unit.

The number 2 is a positive integer. It is located 2 units to the right of zero.

The number 5 is a positive integer. It is located 5 units to the right of zero.

The number −4 is a negative integer. It is located 4 units to the left of zero.

The number −1 is a negative integer. It is located 1 unit to the left of zero.

Locate the numbers on the number line.

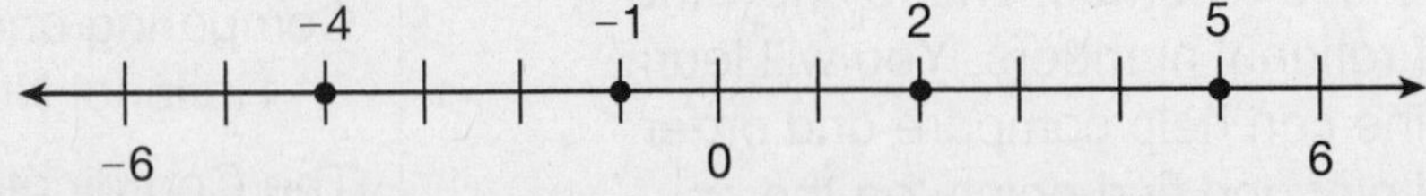

The **opposite** of a number is the number that is the same distance from 0 on a number line, but on the opposite side of 0. The opposite of a positive number is a negative number, and the opposite of a negative number is a positive number. For example, 4 and −4 are opposites.

TIP: Zero is its own opposite.

CCSS: 6.NS.5, 6.NS.6.a

Integers can be used to represent real-world situations. Some keywords that indicate positive integers are *gained, increased, rose, above, more,* and *up*. Some keywords that indicate negative integers are *lost, decreased, dropped, below, less,* and *down*.

Example

What integer is represented by the **bold words** in the following sentence?

The lowest temperature ever recorded in the United States was **80 degrees Fahrenheit below 0**.

The integer that represents **80 degrees Fahrenheit below 0** is **−80**.

Example

A baby boy typically experiences a **growth of 10 inches** during his first year of life. Explain what the 0 means in the situation.

The integer that represents a **growth of 10 inches** is **10.** In this situation, the 0 represents the height of the baby boy at birth.

Example

What integer is represented by the **bold words** in the following sentence? Explain what the 0 means in the situation.

A scuba diver dives **30 meters below the ocean's surface**.

The integer that represents **30 meters below the ocean's surface** is **−30.** In this situation, the 0 represents the surface of the ocean.

Practice

Directions: For questions 1 and 2, write the numbers in their correct place on the number line.

1. 7, −6, −9, 3

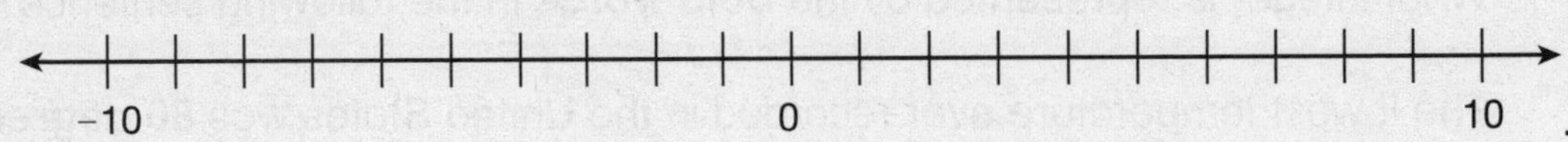

2. −12, 3, 18, −19, −3

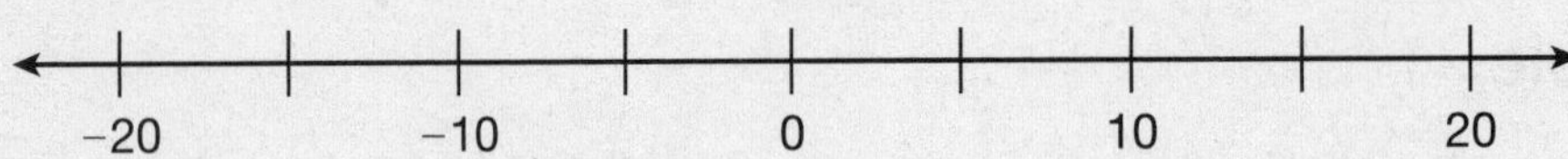

Directions: For questions 3 through 8, use the number line below. Write the integer that belongs at each point.

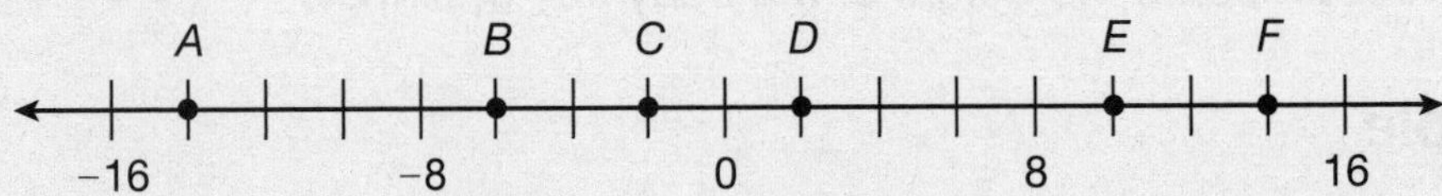

3. *A* __________

4. *B* __________

5. *C* __________

6. *D* __________

7. *E* __________

8. *F* __________

Directions: For questions 9 through 16, write the opposite of the number.

9. −6 __________

10. 11 __________

11. 15 __________

12. 77 __________

13. −80 __________

14. −12 __________

15. 1202 __________

16. −508 __________

CCSS: 6.NS.5, 6.NS.6.a

Directions: For questions 17 through 21, write the integer that is represented by the **bold words in each sentence.**

17. Owen's kitten weighs **4 ounces more** this month than it did last month. ________

18. The height of the skyscraper **increased 25 feet** during construction last week.

    ____

19. The amount in Beth's bank account **increased 195 dollars.** ________

20. A submarine is **400 feet below** the surface of the water. ________

21. Antoine hit **8 fewer home runs** than he did last year. ________

Directions: For questions 22 through 25, write the integer that is represented by the **bold words in each sentence.** Then, interpret what the 0 means in the situation.

22. Holly earned **$800 more** than she did last year. ________

    ____

23. A desktop computer has a mass of **2 kilograms more** than a notebook computer.

    ____

    ____

24. The temperature of a patient **decreased by 2 degrees Celsius.** ________

    ____

25. Isabella is **7 centimeters shorter** than her brother Lorenzo. ________

    ____

    Explain how you know whether this integer is negative or positive.

    ____

Lesson 2: Rational Numbers

A **rational number** is any number that can be expressed as a fraction in the form $\frac{a}{b}$. For example, $-\frac{2}{5}$, $\frac{1}{8}$, and $\frac{20}{1}$ are all rational numbers. You can write and find rational numbers on a horizontal number line.

Example

Write -2, $1\frac{1}{5}$, and $-\frac{4}{5}$ in their correct places on the number line.

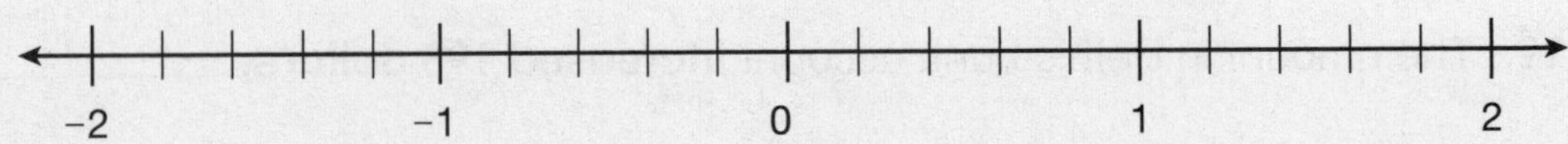

First, you need to figure out what each mark on the number line represents. On the number line above, there are 5 marks from one integer to the next. Therefore, each mark represents $\frac{1}{5}$. You can consider each mark as $\frac{1}{5}$ greater than the mark to the left. See below.

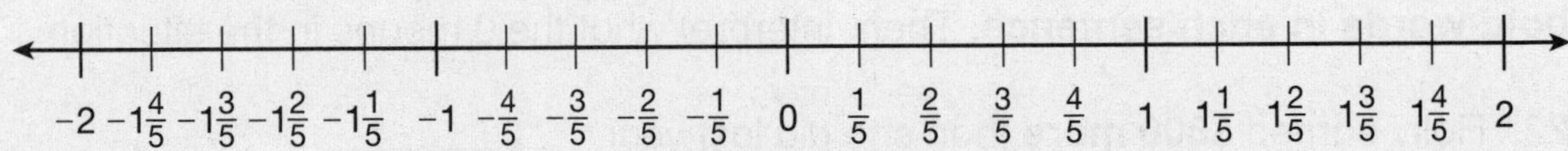

Now you can write -2, $1\frac{1}{5}$, and $-\frac{4}{5}$ on the number line.

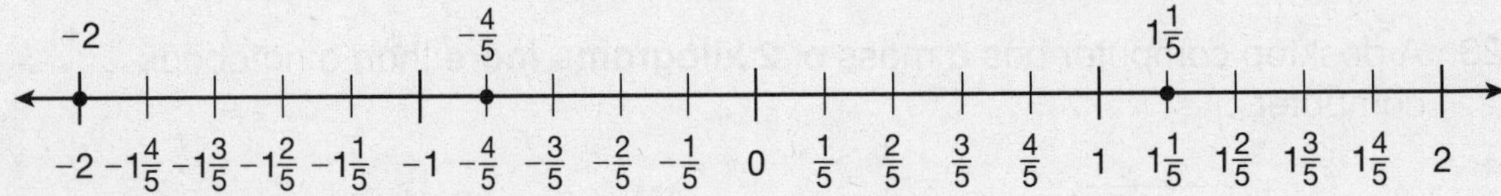

Example

What numbers are represented by *A, B, C,* and *D* on the following number line?

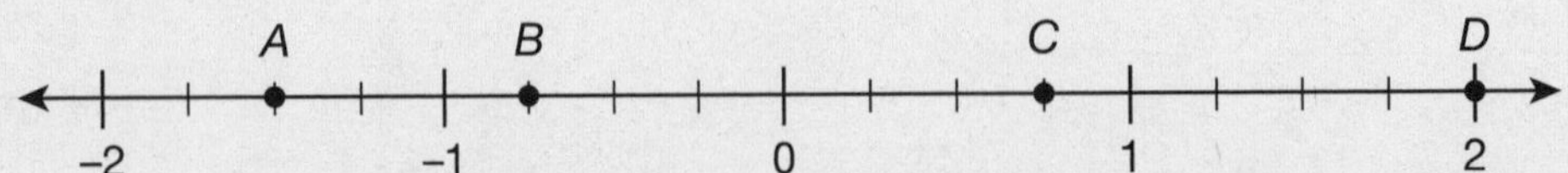

There are 4 tick marks from one integer to the next. Therefore, each mark represents $\frac{1}{4}$. Now you can figure out what rational numbers *A, B, C,* and *D* represent.

A: $-1\frac{2}{4}$ or $-1\frac{1}{2}$ *B:* $-\frac{3}{4}$ *C:* $\frac{3}{4}$ *D:* 2

CCSS: 6.NS.6.c

You can write and find rational numbers on a vertical number line.

Example

Write -2, $-1\frac{1}{2}$, and $\frac{1}{2}$ in their correct places on the number line.

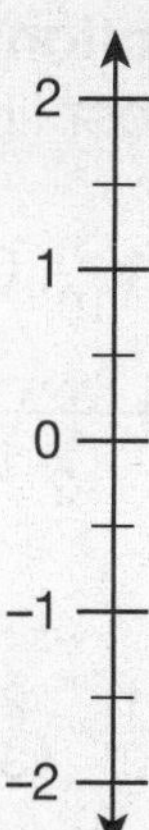

First, you need to figure out what each mark on the number line represents. On the number line to the right, there are 2 marks from one integer to the next. Therefore, each mark represents $\frac{1}{2}$. You can consider each mark as $\frac{1}{2}$ greater than the mark below it.

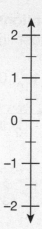

Now you can write -2, $-1\frac{1}{2}$, and $\frac{1}{2}$ on the number line.

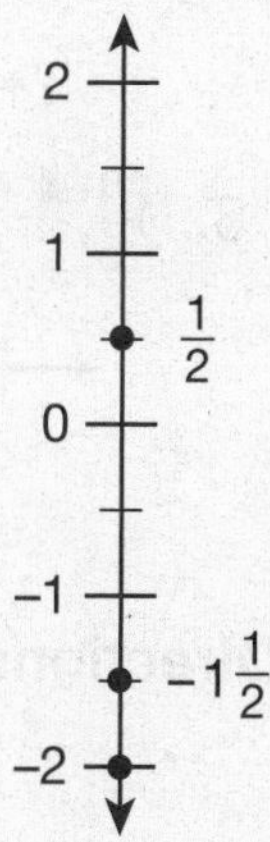

Example

What numbers are represented by *A, B, C,* and *D* on the number line to the right?

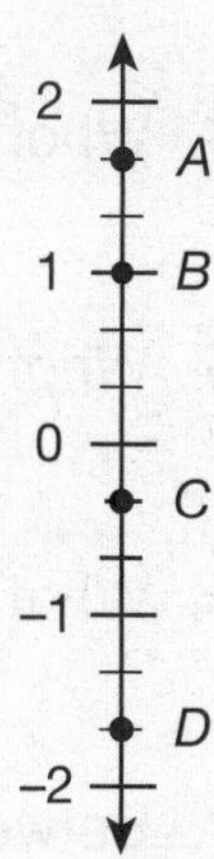

There are 3 tick marks from one integer to the next. Therefore, each mark represents $\frac{1}{3}$. Now you can figure out what rational numbers *A, B, C,* and *D* represent.

A: $1\frac{2}{3}$ *B*: 1 *C*: $-\frac{1}{3}$ *D*: $-1\frac{2}{3}$

Practice

Directions: For questions 1 through 3, write the numbers in their correct places on the number line.

1. $1\frac{3}{10}$, 0, $-\frac{3}{5}$

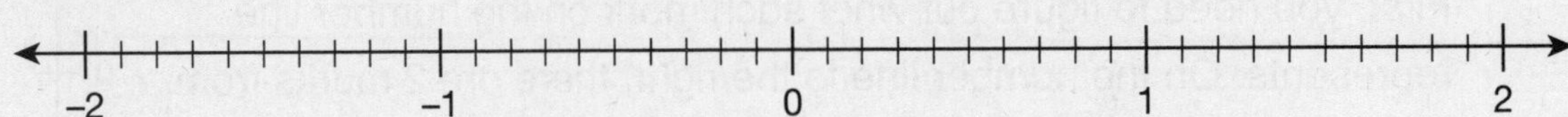

2. $\frac{1}{3}$, $-\frac{2}{3}$, 1

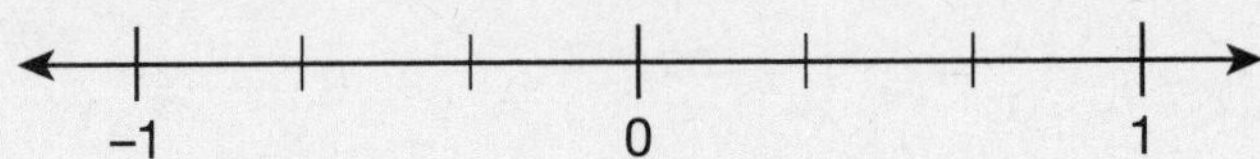

3. $1\frac{4}{5}$, $-\frac{1}{5}$, 1

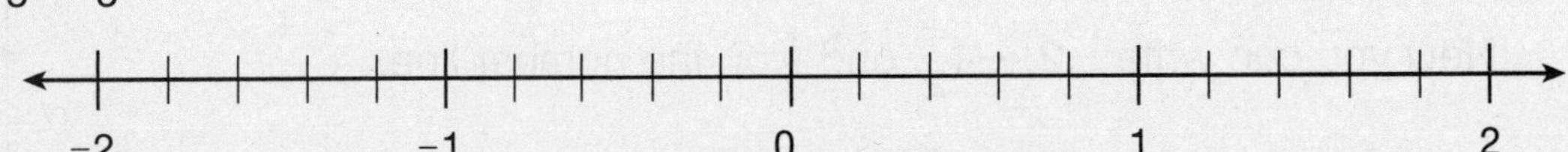

Directions: Use the following number line to answer questions 4 through 8.

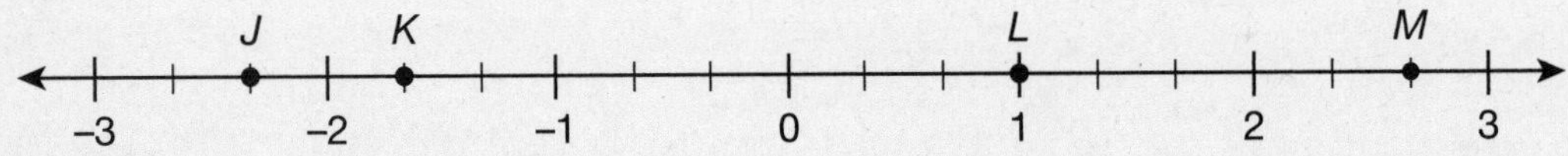

4. What point is at $-1\frac{2}{3}$? ____________

5. What rational number is at *J*? ____________

6. What rational number is at *L*? ____________

7. Between which two points would $\frac{1}{3}$ be? ____________

8. Between which two points would 2 be? ____________

CCSS: 6.NS.6.c

Directions: For questions 9 and 10, write the numbers in their correct places on the number line.

9. $-1\frac{1}{4}, \frac{3}{4}, 2, -\frac{2}{4}$

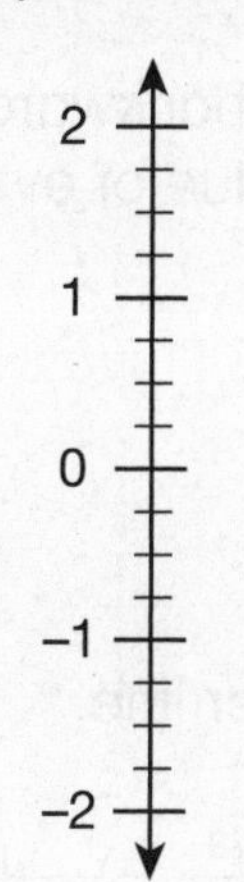

10. $3, -\frac{1}{2}, -2, 1\frac{1}{2}$

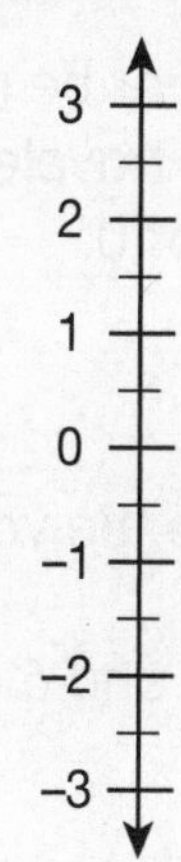

Directions: Use the following number line to answer questions 11 through 15.

11. What point is at $-\frac{2}{3}$? ____________

12. What rational number is at *Q*? ____________

13. What rational number is at *S*? ____________

14. Between which two points would 0 be? ____________

15. Between which two points would $-1\frac{2}{3}$ be? ____________

Explain how you found the location of $-1\frac{2}{3}$ on the number line.

____

____

____

3
P
2
1
Q
0
R
–1
–2 *S*
–3

Lesson 3: Absolute Value

The **absolute value** of a number is that number's distance from 0 on a number line. When you write the absolute value of a number *n*, use the notation $|n|$.

A distance can never be negative. For example, even walking backward for 20 feet still means that you've traveled 20 feet. Therefore, the absolute value of every number will be either positive or 0.

Example

What are the values of $|-6|$ and $|6|$?

Both -6 and 6 are exactly 6 units from 0 on a number line.

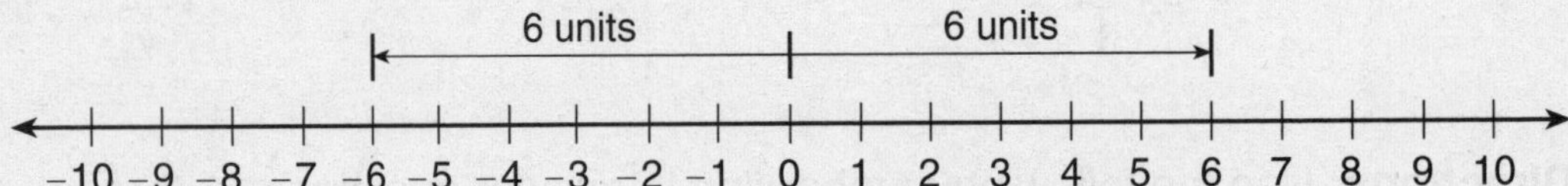

The absolute value of $-6 = |-6| = 6$. The absolute value of $6 = |6| = 6$.

Example

What is the value of $\left|-\frac{3}{4}\right|$?

$-\frac{3}{4}$ is $\frac{3}{4}$ units from 0 on a number line.

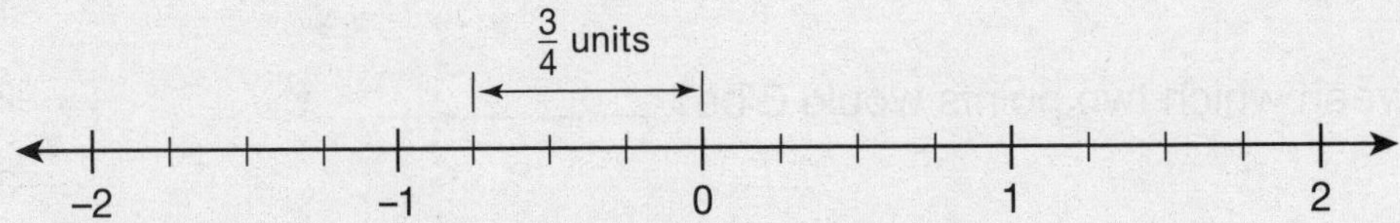

The absolute value of $-\frac{3}{4} = \left|-\frac{3}{4}\right| = \frac{3}{4}$.

CCSS: 6.NS.7.c, 6.NS.7.d

Absolute value can be used to represent a quantity in a real-world situation.

Example

A fisherman's hook hangs 12 feet under the surface of the water. Write an expression using absolute value to interpret the distance that the hook hangs from the surface of the water.

The word *under* means the integer should be negative. The integer can be written as -12. However, distance cannot be negative. Therefore, the distance can be represented as the absolute value of -12.

$|-12| = 12$

Absolute value can also be used to compare quantities.

Example

The following table shows two test scores for four students in Ms. Yan's class, as well as the change in score between the two tests.

Student	Test 1 Score	Test 2 Score	Change from Test 1 to Test 2
Amelia	93	97	4
Bobby	88	82	−6
Cathy	92	89	−3
Denzel	89	94	5

Which student's test score changed the most between test 1 and test 2?

To answer this question, you need to look at the numbers that show the change from test 1 to test 2. You then must find the number that is the farthest from 0 on the number line. The 0 does not represent a score of 0, however. The 0 represents the score on test 1. The following number line shows the four absolute values.

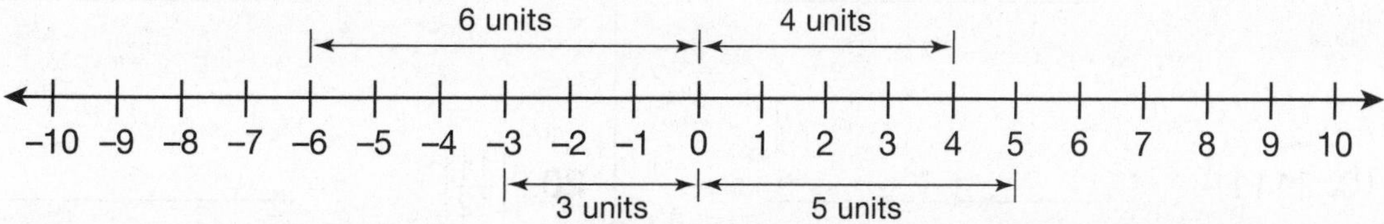

The -6 change from test 1 to test 2 represents the score that changed the most. Bobby's test score changed the most between test 1 and test 2.

Practice

Directions: For questions 1 through 20, determine each value.

1. $|-7|$ ____________
2. $|14|$ ____________
3. $|8|$ ____________
4. $|-17|$ ____________
5. $|0|$ ____________
6. $|-1|$ ____________
7. $|32|$ ____________
8. $|-9|$ ____________
9. $|-20|$ ____________
10. $|111|$ ____________
11. $|-15|$ ____________
12. $|28|$ ____________
13. $\left|\frac{1}{5}\right|$ ____________
14. $|-78|$ ____________
15. $\left|-\frac{2}{3}\right|$ ____________
16. $|40|$ ____________
17. $\left|-\frac{7}{8}\right|$ ____________
18. $|-88|$ ____________
19. $|401|$ ____________
20. $\left|\frac{11}{12}\right|$ ____________

CCSS: 6.NS.7.c, 6.NS.7.d

Directions: For questions 21 through 24, determine a value that fits the real-world situation.

21. The altitude of a hot air balloon changed by −25 feet in a minute while preparing to land. How many feet did the balloon travel during the minute?

____

22. A mountain climber traveled 500 feet up a mountain during an afternoon. By how many feet did the climber's elevation change during the afternoon?

____

23. The average temperature of the ocean from September to October changed by −4°F. What was the overall change in temperature of the water in degrees Fahrenheit?

____

24. The balance in Jason's bank account is listed as −135 dollars. What is the amount of his debt to the bank?

____

25. The following table shows the population change from 2005 to 2010 for four species at a local zoo.

Species	2005	2010	Change from 2005 to 2010
Polar Bear	11	12	1
Emperor Penguin	17	13	−4
Tiger	1	1	0
Chimpanzee	18	21	3

Which species' population changed the most between 2005 and 2010?

____

Explain how you found your answer.

____

____

____

Lesson 4: Comparing and Ordering Rational Numbers

Rational numbers can be compared using a number line. As you move from left to right on a horizontal number line, or from bottom to top on a vertical number line, the numbers are ordered from least to greatest. All numbers to the left of 0 or below 0 will be negative. You can compare or order numbers using the less than symbol, $<$, or the greater than symbol, $>$.

The number line below shows that $-2\frac{1}{2}$ is to the left of -1. Therefore, $-2\frac{1}{2}$ is less than -1. You can represent this inequality with the less than symbol, $<$: $-2\frac{1}{2} < -1$. You can also represent this inequality with the greater than symbol, $>$: $-1 > -2\frac{1}{2}$.

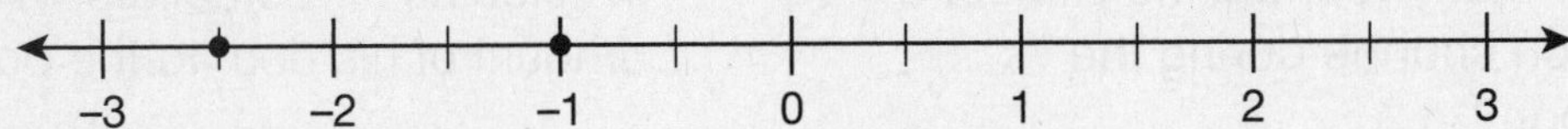

Example

Write the following numbers in order from **greatest** to **least**. Then compare them using $<$, $>$, or $=$.

12 -13 -35 25

First, find the position of the numbers on a number line.

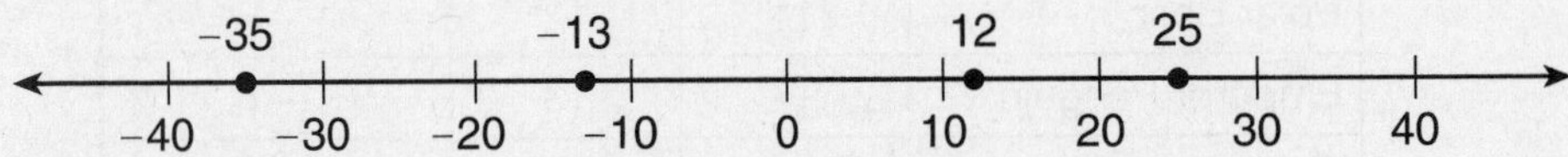

Then, look at where the numbers are located in relation to each other on the number line. 25 is the farthest to the right. It is therefore the greatest number. Moving left, the next greatest number is 12. The next number to the left is -13, then -35.

The numbers in order from **greatest** to **least** are 25, 12, -13, and -35.

Now that the order of the numbers is known, use the comparison symbols to compare the numbers: $25 > 12 > -13 > -35$. This means that 25 is to the right of 12, 12 is to the right of -13, and -13 is to the right of -35.

You can also compare and order rational numbers without a number line. To compare fractions with the same denominator, compare the numerators. To compare fractions with different denominators, you need to multiply each fraction by the other fraction's denominator. Then the denominators will be the same and you can compare the numerators.

Example

Compare the rational numbers and order them from **greatest** to **least.** Then compare them using $<$, $>$, or $=$.

$\frac{1}{8}$ $\frac{7}{8}$ $-\frac{3}{8}$ $\frac{5}{8}$

Because the fractions have the same denominator, you can compare the numerators. The numerators are 1, 7, -3, and 5. The fraction with the greatest numerator is the greatest fraction. 7 is the largest number, so $\frac{7}{8}$ is the largest fraction. The next greatest numerator is 5, so $\frac{5}{8}$ is the next largest fraction. -3 is less than 0, so 1 is greater than -3.

In order from greatest to least, the fractions are $\frac{7}{8}$, $\frac{5}{8}$, $\frac{1}{8}$, and $-\frac{3}{8}$

Now that the order of the numbers is known, use the comparison symbols to compare the numbers: $\frac{7}{8} > \frac{5}{8} > \frac{1}{8} > -\frac{3}{8}$.

Example

Compare the numbers from **least** to **greatest** using $<$, $>$, or $=$.

$\frac{4}{5}$ $\frac{3}{4}$

The fractions have different denominators, so multiply each fraction by the other fraction's denominator.

$\frac{4}{5} = \frac{4 \times 4}{5 \times 4} = \frac{16}{20}$ $\frac{3}{4} = \frac{3 \times 5}{4 \times 5} = \frac{15}{20}$

Now that the fractions have the same denominator, you can compare the numerators. Because 16 is greater than 15, $\frac{4}{5}$ is greater than $\frac{3}{4}$.

$\frac{4}{5} > \frac{3}{4}$

TIP: The top number in a fraction is the numerator. The bottom number is the denominator. One way to remember those terms is to think of a word that starts with N, like **n**umerator, that makes you think of the top, such as **N**orth Pole. Then think of a word that starts with D, like **d**enominator, that makes you think of the bottom, such as **d**own.

Example

The lowest recorded temperature in Florida was −2°F. The lowest recorded temperature in Mississippi was −19°F. The lowest recorded temperature in Hawaii was 12°F. The lowest recorded temperature in Louisiana was −16°F. Compare the low temperatures and order them using $<$, $>$, or $=$.

To compare the temperatures, you can plot them on a number line.

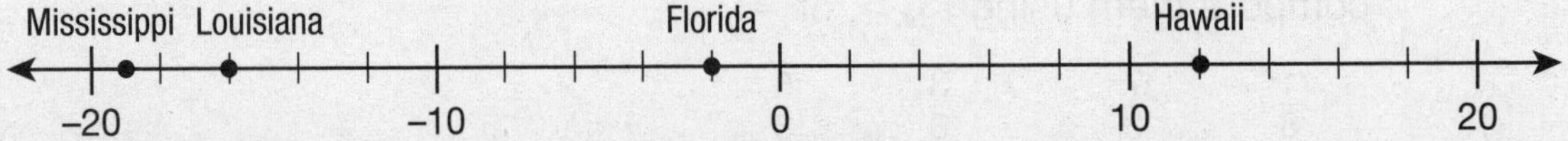

The numbers in order from **least** to **greatest** are −19 for Mississippi, −16 for Louisiana, −2 for Florida, and then 12 for Hawaii. Now that the order of the numbers is known, use the comparison symbols to compare the numbers: $-19 < -16 < -2 < 12$.

Example

A supermarket sells two packages of rice. Package A is listed as $\frac{2}{3}$ pounds. Package B is listed as $\frac{3}{5}$ pounds. Compare the sizes of the packages using the comparison symbols $<$ or $>$.

The fractions have different denominators, so multiply both fractions by the other fraction's denominator.

$$\frac{2}{3} = \frac{2 \times 5}{3 \times 5} = \frac{10}{15} \qquad \frac{3}{5} = \frac{3 \times 3}{5 \times 3} = \frac{9}{15}$$

Now that the fractions have the same denominator, you can compare the numerators. Because 10 is greater than 9, $\frac{2}{3}$ is greater than $\frac{3}{5}$: $\frac{2}{3} > \frac{3}{5}$.

Practice

Directions: For questions 1 through 6, refer to the number line, and use $<$, $>$, or $=$ to compare the integers.

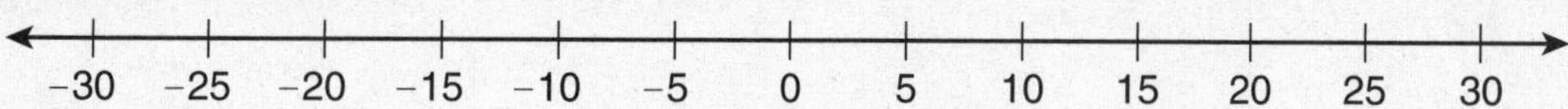

1. −12 ________ 0
2. −4 ________ −16
3. −30 ________ −30
4. −2 ________ 2
5. −18 ________ −13
6. 27 ________ 9

Directions: For questions 7 through 12, use $<$, $>$, or $=$ to compare the integers.

7. −126 ______________ −216
8. −412 ______________ 398
9. 1927 ______________ 832
10. −57 ______________ 65
11. −59 ______________ −59
12. 0 ______________ −105

Directions: For questions 13 and 14, write the integers in order from **least** to **greatest.**

13. −25, 25, 7, 0, −10, −16, −5, 1, −2, −1, 4, 5

____

14. 3, 8, 1, −11, −9, 0, −13, −8, 11, −7

____

15. Is the following list of integers in order from **greatest** to **least** or from **least** to **greatest**?

    −10, −9, −8, −6, −5

____

Directions: For questions 16 through 20, use the number line to help you order the numbers from **least** to **greatest**.

16. $\frac{1}{4}$ $-\frac{3}{4}$ 0 1 ____________________

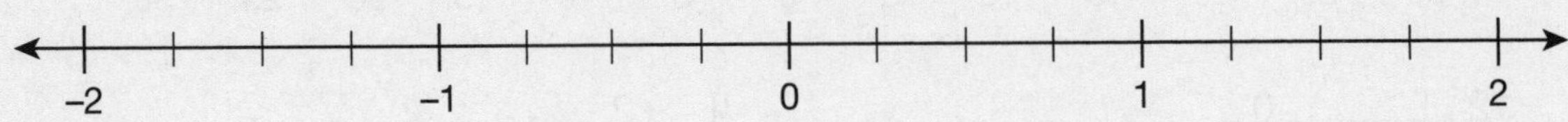

17. 2 $-1\frac{2}{4}$ $-\frac{1}{4}$ $\frac{1}{2}$ ____________________

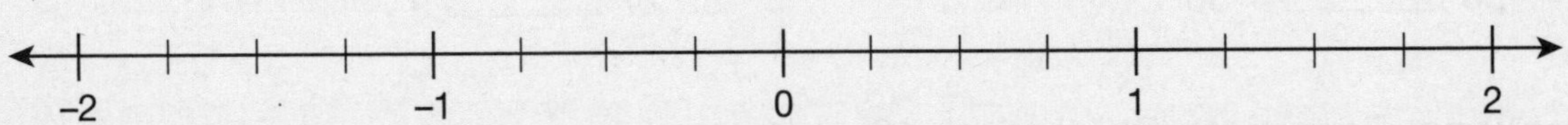

18. $1\frac{4}{5}$ –2 $\frac{2}{5}$ $-\frac{3}{5}$ ____________________

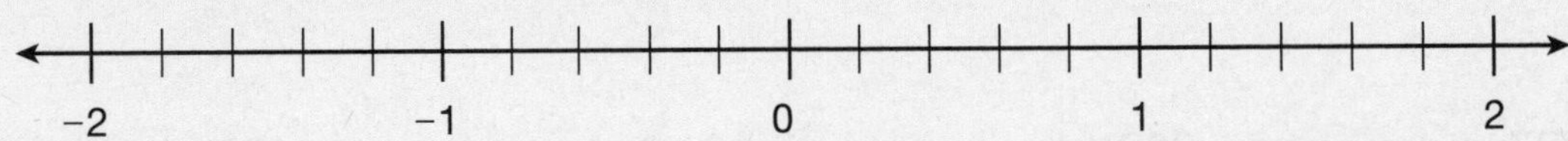

19. $1\frac{1}{5}$ $1\frac{2}{5}$ –1 $-\frac{1}{5}$ ____________________

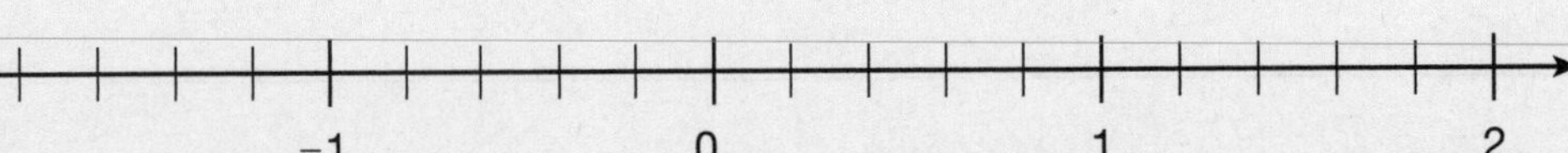

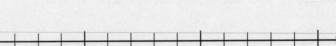

20. $1\frac{3}{5}$ $-1\frac{3}{5}$ $1\frac{2}{5}$ $-1\frac{2}{5}$ ____________________

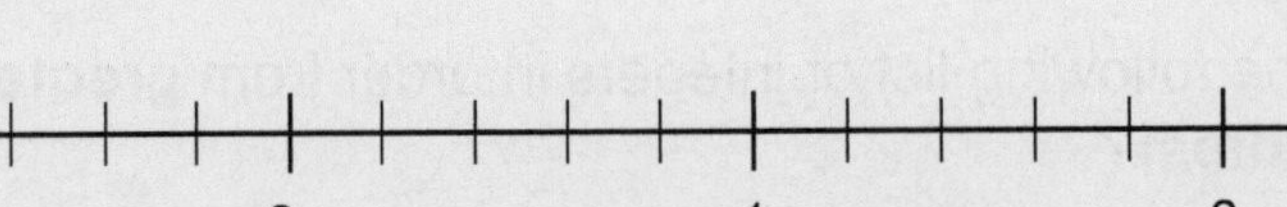

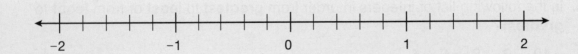

21. A businesswoman invested in four different stocks. The amounts that each stock earned for her are listed in the following table.

Stock	Return of Investment (in $)
Macrotechnico	−200
Sun Solar Systems	880
Benefact & Co.	−95
Posada Pizza	125

List the order of the stocks from the **least** return on her investment to the **greatest** return on her investment.

____

22. The record low temperatures for Portland, ME, for the first four months of the year, are shown in the table below.

Month	Record Low Temperature (in °F)
January	−26
February	−39
March	−21
April	8

List the order of the months from the **least** temperature to the **greatest** temperature.

____

23. A recipe calls for $\frac{3}{4}$ cup of flour and $\frac{5}{8}$ cup of sugar. Compare the amounts of flour and sugar using $<$, $>$, or $=$.

____

Explain how you were able to compare the fractions.

____

____

____

Lesson 5: The Coordinate Plane

The **coordinate plane** is a system of two number lines. The ***x*-axis** is the horizontal number line in a coordinate plane. The ***y*-axis** is the vertical number line in a coordinate plane.

The location of a point on the coordinate plane can be described by its distance along both number lines. An **ordered pair** is a pair of numbers (*x*, *y*) used to locate a point on a coordinate plane. The ***x*-coordinate** is the first number in an ordered pair. The ***x*-coordinate** describes the distance left or right from 0 on the *x*-axis. The ***y*-coordinate** is the second number in an ordered pair. The ***y*-coordinate** describes the distance up or down from 0 on the *y*-axis.

Example

Identify the location of the points on the coordinate plane.

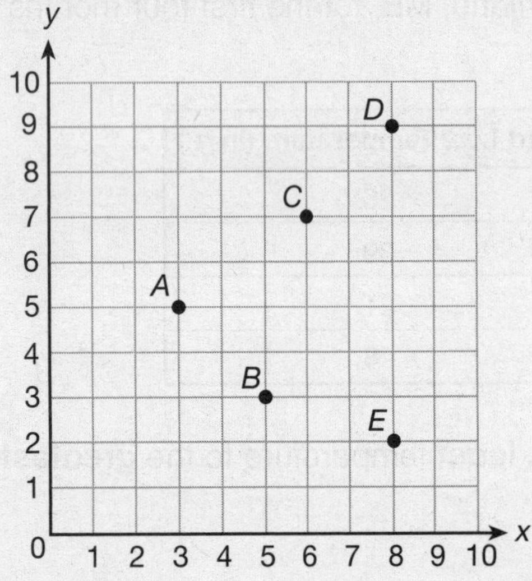

To locate point *A*, count along the *x*-axis. The *x*-coordinate of point *A* is 3. Then count up the *y*-axis. The *y*-coordinate of point *A* is 5. Point *A* is located at (3, 5).

To locate the remaining points, do the same for each point.

Point *B* is located at (5, 3).

Point *C* is located at (6, 7).

Point *D* is located at (8, 9).

Point *E* is located at (8, 2).

CCSS: 6.NS.6.b, 6.NS.6.c

The coordinates of the point where the number lines intersect are (0, 0). This point is called the **origin.** However, the number lines in a coordinate plane can extend below and to the left of 0. Positive *x*-coordinates are to the right of the origin, and negative *x*-coordinates are to the left of the origin. Positive *y*-coordinates are above the origin, and negative *y*-coordinates are below the origin.

When the axes are extended in both directions, they divide the coordinate plane into four parts, also known as **quadrants.**

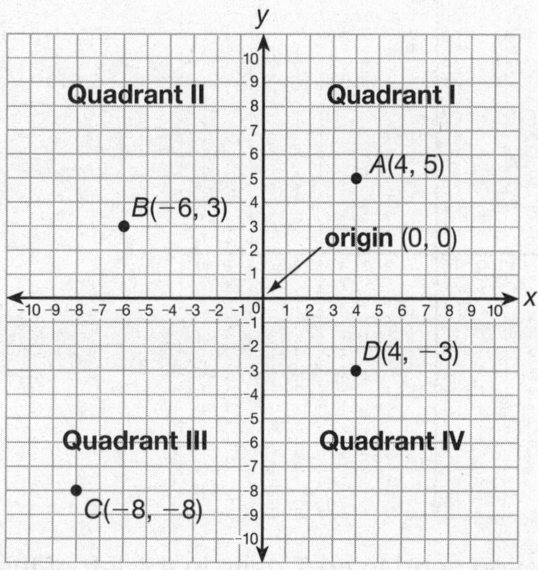

Example

Identify the location of the points on the coordinate plane.

Point *A* is located in Quadrant I. All the points in Quadrant I have two positive numbers in their ordered pair, (*x*, *y*). Point *A* is located at (4, 5).

Points in Quadrants II, III, and IV follow these general patterns; Quadrant II: (−*x*, *y*); Quadrant III: (−*x*, −*y*); Quadrant IV: (*x*, −*y*). You can use these patterns to check if your points have the right coordinates.

Point *B* is in Quadrant II: (−6, 3)

Point *C* is in Quadrant III: (−8, −8)

Point *D* is in Quadrant IV: (4, −3)

You can use the coordinate plane to plot points. The ordered pairs for the points can include any rational numbers.

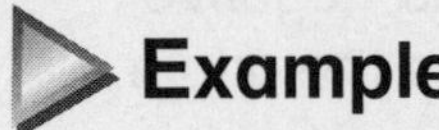

Example

Plot the following points on the coordinate plane.

$A\,(-7, 5)$ $\quad$ $B\,(4, -8\frac{1}{2})$

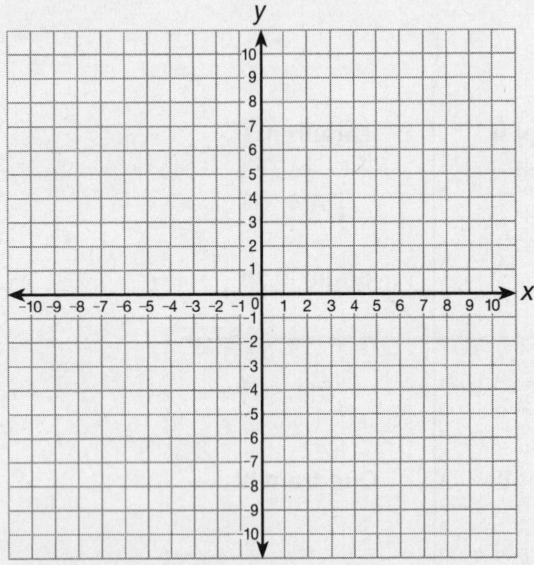

To plot point *A,* you need to go 7 units to the left of the origin along the *x*-axis. Then you need to go 5 units up along the *y*-axis.

To plot point *B,* you need to go 4 units to the right of the origin along the *x*-axis. Then you need to go $8\frac{1}{2}$ units down along the *y*-axis. That will be halfway between the −8 and −9 markers along the *y*-axis.

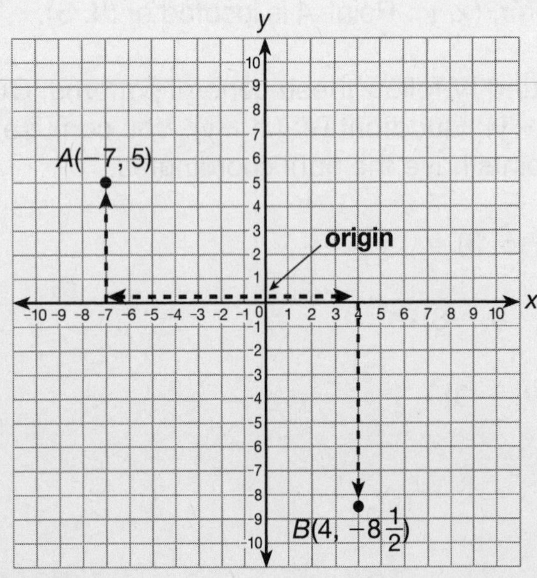

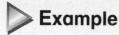

CCSS: 6.NS.6.b, 6.NS.6.c

If the ordered pairs of two points are different by a negative symbol, the points are reflections over an axis. If the ordered pairs of the points are different by two negative symbols, the points are reflections over both axes.

Example

Plot the following three pairs of points on a coordinate plane. Then compare their relationships with the axes.

A (5, 7) and *B* (−5, 7)
C (−2, 4) and *D* (−2, −4)
E (3, 8) and *F* (−3, −8)

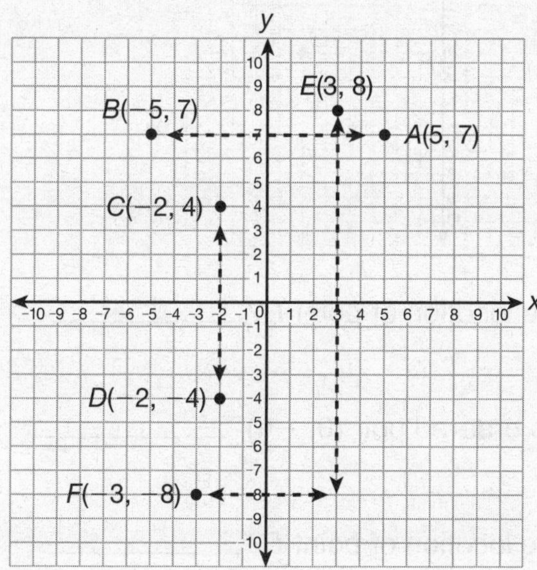

Points *A* and *B* are different because of a negative symbol in the *x*-coordinate. They are a reflection of each other across the *y*-axis.

Points *C* and *D* are different because of a negative symbol in the *y*-coordinate. They are a reflection of each other across the *x*-axis.

Points *E* and *F* are different because of a negative symbol in both the *x*- and the *y*-coordinates. They are a reflection of each other across the *y*-axis and the *x*-axis.

Practice

Directions: Use the coordinate plane below to answer questions 1 through 7.

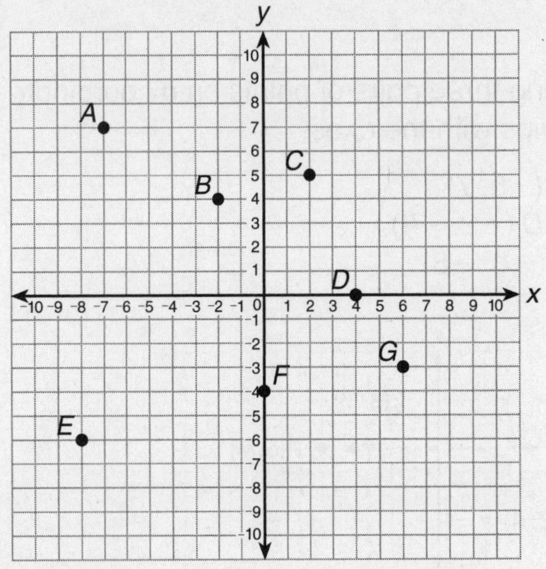

1. What ordered pair represents the location of point *D*? ____________

2. What point is represented by the ordered pair (0, −4)? ____________

3. What ordered pair represents the location of point *B*? ____________

4. What point is represented by the ordered pair (6, −3)? ____________

5. What ordered pair represents the location of point *A*? ____________

6. What point is represented by the ordered pair (2, 5)? ____________

7. What ordered pair represents the location of point *E*? ____________

CCSS: 6.NS.6.b, 6.NS.6.c

Directions: Use the coordinate plane below to answer questions 8 through 15.

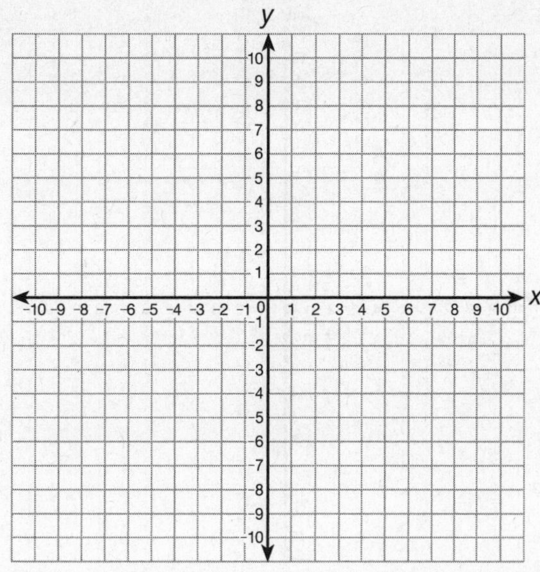

8. Plot point A at $(-4, 5)$.

9. Plot point B at $(3, 9)$.

10. Plot point C at $(-3, 0)$.

11. Plot point D at $(7, -8)$.

12. Plot point E at $(1, 6)$.

13. Plot point F at $(-2, -10)$.

14. Plot point G at $(-5, -5)$.

15. Plot point H at $(5, -5)$.

Directions: Use the coordinate plane below to answer questions 16 through 20.

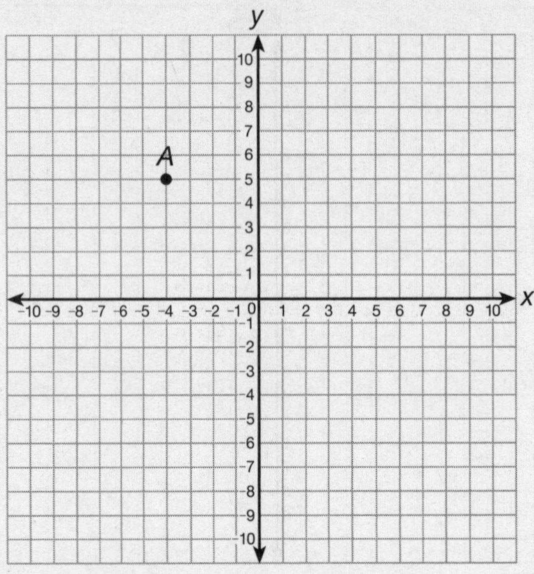

16. Identify the coordinates of point *A*. ____________

17. Plot point *J* at $(7, 8\frac{1}{2})$.

18. Point *B* is a reflection of point *A* over the *y*-axis. Identify the coordinates of point *B*, and then plot the point on the coordinate plane.

____

19. Point *K* has the same coordinates as point *J* except the sign of the *y*-coordinate is opposite. Identify the coordinates of point *K*, and then plot the point.

____

20. Point *C* has the same coordinates as point *A*, but the signs of the *x*- and *y*-coordinates are reversed. Identify the coordinates of point C, and then plot the point.

____

21. Explain how you are able to find point *C* on the coordinate plane without first identifying its coordinates.

____

____

CCSS: 6.NS.2

Lesson 6: Dividing Whole Numbers

When you divide, the number you are dividing is the **dividend**, the number you are using to divide is the **divisor**, and the answer is the **quotient**. Anything that is left over is the **remainder (R)**.

Example

Aubrey has 419 beads. She wants to divide the beads equally among 16 bracelets that she is making. How many beads will Aubrey use for each bracelet?

```
          ┌──────── 1. How many 16s are in 4? 0 Move to the next place value.
          │ ┌────── 2. How many 16s are in 41? 2
          ↓ ↓ ↓──── 5. How many 16s are in 99? 6
      26
  16)419
   − 32↓     ← 3. Multiply 2 and 16; then subtract.
      99     ← 4. Bring down the 9.
    − 96     ← 6. Multiply 6 and 16; then subtract.
       3     ← 7. Remainder (must be less than the divisor)
```

Aubrey will use 26 beads for each bracelet. There will be 3 beads left over. The quotient is therefore 26 R3.

Multiplication and division are inverse operations. To check a division problem, multiply the whole number portion of the quotient by the divisor, and then add the remainder. This should give you the dividend.

Example

Check the quotient and remainder of the previous example.

```
     26     ← Whole number portion of the quotient
   × 16     ← Divisor
    156
  + 260
    416
  +   3     ← Add the remainder.
    419
```

The quotient of 26 R3 is correct.

TIP: To check a multiplication problem, divide the product by one of the factors. This should give you the other factor.

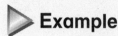

Division can be represented using the division symbol ÷. Always remember that the dividend is the number before the symbol, and the divisor is the number after the symbol.

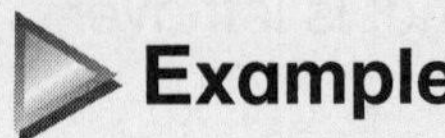

Example

Find the quotient of 545 ÷ 23.

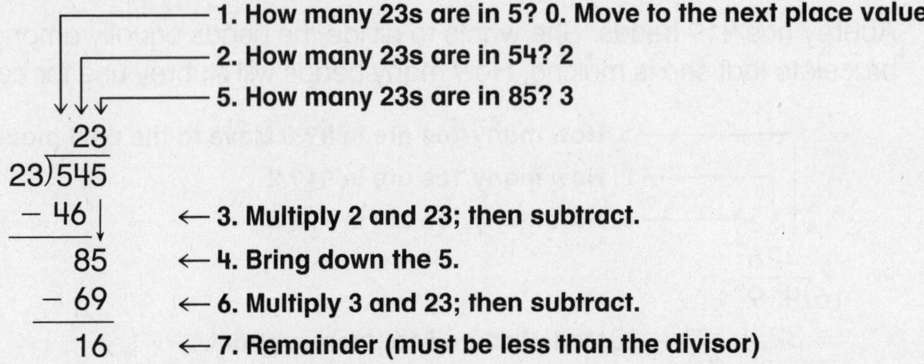
1. How many 23s are in 5? 0. Move to the next place value.
2. How many 23s are in 54? 2
5. How many 23s are in 85? 3
23
23)545
− 46
85
− 69
16
← 3. Multiply 2 and 23; then subtract.
← 4. Bring down the 5.
← 6. Multiply 3 and 23; then subtract.
← 7. Remainder (must be less than the divisor)

The quotient of 545 ÷ 23 is 23 R16.

It's always a good idea to check the answer to a division problem using multiplication.

Example

Check the quotient and remainder of the previous example.

23 ← **Whole number portion of the quotient**
× 23 ← **Divisor**
69
+ 460
529
+ 16 ← **Add the remainder.**
545

The quotient of 23 R16 is correct.

CCSS: 6.NS.2

Practice

Directions: For questions 1 through 8, divide. Write your answer with a remainder, if necessary.

1. A baker bakes 752 muffins in a weekend. She sells the muffins in packs of 12. How many packs of muffins did she sell, and how many did she have left over?

   ____

2. $414 \div 28 =$ ____________

3. $13\overline{)888}$

4. $703 \div 37 =$ ____________

5. A book publisher prints 909 copies of a book. He packs the books into boxes with exactly 18 books in each box. How many boxes will he be able to fill, and how many books will be left over?

   ____

6. $41\overline{)862}$

7. $15\overline{)520}$

8. $992 \div 31 =$ ____________

9. A teacher has a stack of 132 worksheets. She wants to give the same number of worksheets to each of her students. If there are 24 students in her class, how many worksheets will each student get? How many will be left over?

   ____

   Explain how you found your answer.

   ____

   ____

   ____

Lesson 7: Common Factors and Multiples

A **multiple** of a whole number is found by multiplying that number by any other whole number.

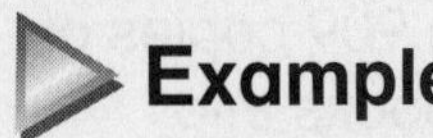

Example

What are the first ten multiples of 4?

4 × 1 = **4**	4 × 6 = **24**
4 × 2 = **8**	4 × 7 = **28**
4 × 3 = **12**	4 × 8 = **32**
4 × 4 = **16**	4 × 9 = **36**
4 × 5 = **20**	4 × 10 = **40**

The first ten multiples of 4 are 4, 8, 12, 16, 20, 24, 28, 32, 36, and 40.

A number that is a multiple of two or more numbers is a **common multiple** of those numbers. Zero is not considered a common multiple. The smallest number that is a common multiple of a set of numbers is called the **least common multiple (LCM)** of that set of numbers.

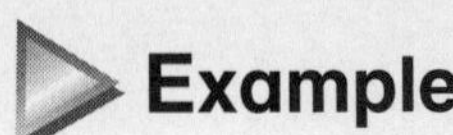

Example

What is the LCM of 4 and 6?

The first ten multiples of 4 are listed in the previous example. The first ten multiples of 6 are as follows:

6 × 1 = **6**	6 × 6 = **36**
6 × 2 = **12**	6 × 7 = **42**
6 × 3 = **18**	6 × 8 = **48**
6 × 4 = **24**	6 × 9 = **54**
6 × 5 = **30**	6 × 10 = **60**

The first ten multiples of 6 are 6, 12, 18, 24, 30, 36, 42, 48, 54, and 60.

Common multiples of 4 and 6 include 12, 24, 36, and 48.

The LCM of 4 and 6 is 12.

TIP: There are an infinite number of multiples for any number or common multiples for any two numbers. However, there will only be one *least* common multiple.

CCSS: 6.NS.4

A **factor** of a whole number is any whole number that divides the first number evenly (with no remainder). A number is **divisible** by its factors because they divide evenly into the number. The factors of a number are less than or equal to the number.

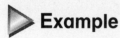

Example

What are the factors of 15?

$15 \div \mathbf{1} = 15$	$15 \div 6 = 2\text{ R}3$	$15 \div 11 = 1\text{ R}4$
$15 \div 2 = 7\text{ R}1$	$15 \div 7 = 2\text{ R}1$	$15 \div 12 = 1\text{ R}3$
$15 \div \mathbf{3} = 5$	$15 \div 8 = 1\text{ R}7$	$15 \div 13 = 1\text{ R}2$
$15 \div 4 = 3\text{ R}3$	$15 \div 9 = 1\text{ R}6$	$15 \div 14 = 1\text{ R}1$
$15 \div \mathbf{5} = 3$	$15 \div 10 = 1\text{ R}5$	$15 \div \mathbf{15} = 1$

The factors of 15 are 1, 3, 5, and 15.

A number that is a factor of two or more numbers is a **common factor** of those numbers. The greatest number that is a common factor is called the **greatest common factor (GCF)**.

Example

What is the GCF of 15 and 9?

The factors of 15 are listed in the previous example. The factors of 9 are as follows:

$9 \div \mathbf{1} = 9$	$9 \div 4 = 2\text{ R}1$	$9 \div 7 = 1\text{ R}2$
$9 \div 2 = 4\text{ R}1$	$9 \div 5 = 1\text{ R}4$	$9 \div 8 = 1\text{ R}1$
$9 \div \mathbf{3} = 3$	$9 \div 6 = 1\text{ R}3$	$9 \div \mathbf{9} = 1$

The factors of 9 are 1, 3, and 9.

The common factors of 15 and 9 are 1 and 3.

The GCF of 15 and 9 is 3.

Practice

Directions: For questions 1 through 6, list the first seven multiples of the given number.

1. 5: ______, ______, ______, ______, ______, ______, ______

2. 6: ______, ______, ______, ______, ______, ______, ______

3. 8: ______, ______, ______, ______, ______, ______, ______

4. 9: ______, ______, ______, ______, ______, ______, ______

5. 10: ______, ______, ______, ______, ______, ______, ______

6. 12: ______, ______, ______, ______, ______, ______, ______

Directions: Use the multiples from questions 1 through 6 to answer questions 7 through 13.

7. List at least two common multiples of 6 and 12. ______

8. List at least two common multiples of 8 and 12. ______

9. List at least two common multiples of 5 and 10. ______

10. What is the LCM of 5 and 6? ______

11. What is the LCM of 6 and 8? ______

12. What is the LCM of 6 and 9? ______

13. What is the LCM of 9 and 12? ______

Directions: For questions 14 through 19, list all the factors of the number.

14. 7 ____________

15. 8 ____________

16. 12 ____________

17. 35 ____________

18. 56 ____________

19. 60 ____________

Directions: Use the factors from 14 through 19 to answer questions 20 through 26.

20. List the common factors of 56 and 60. ____________

21. List the common factors of 8 and 12. ____________

22. List the common factors of 12 and 60. ____________

23. What is the GCF of 8 and 12? ____________

24. What is the GCF of 7 and 35? ____________

25. What is the GCF of 12 and 56? ____________

26. What is the GCF of 7 and 8? ____________

    Explain how you found your answer.

    ____

    ____

    ____

Lesson 8: Adding and Subtracting Decimals

When you compute with decimals, the placement of the decimal point is very important. You must line up the decimal points and place values in the numbers you are adding or subtracting. You may need to use zeros as placeholders. Then add or subtract the two decimals as if they were whole numbers. Finally, move the decimal point straight down into the sum or difference.

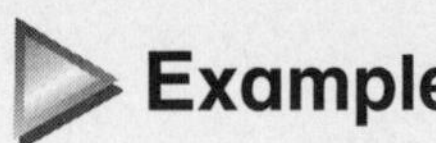

Example

What is the sum of 3.13 + 2.392?

Line up the numbers by their decimal points. You can add a zero as a placeholder at the end of 3.13. That way, both numbers have the same number of digits to the right of the decimal point. Then add.

$$\begin{array}{r} {}^{1} \\ 3.130 \\ +\ 2.392 \\ \hline 5.522 \end{array}$$

3.13 + 2.392 = 5.522

Example

The Mueller family checks two suitcases at the airport. The weight of the larger suitcase is 44.8 pounds. The weight of the lighter suitcase is 31.337 pounds. What is the combined weight of the two suitcases?

Line up the numbers by their decimal points. Add two zeroes to 44.8 as placeholders so both numbers have the same number of digits to the right of the decimal point. Then add.

$$\begin{array}{r} {}^{1} \\ 44.800 \\ +\ 31.337 \\ \hline 76.137 \end{array}$$

The combined weight of the two checked suitcases is 76.137 pounds.

CCSS: 6.NS.3

Example

What is the difference of 8.19 − 4.123?

Line up the numbers by their decimal points. Add a zero as a placeholder at the end of 8.19. That way, both numbers have the same number of digits to the right of the decimal point. Then subtract.

```
     810
  8.1~~9~~~~0~~
− 4.123
———————
  4.067
```

8.19 − 4.123 = 4.067

Example

A quart of blueberries costs $3.99. A quart of organic blueberries costs $5.75. How much more expensive is the quart of organic blueberries?

Line up the numbers by their decimal points. Then subtract.

```
  4 1615
  ~~5~~.~~7~~~~5~~
− 3.99
——————
  1.76
```

The organic blueberries are $1.76 more expensive.

Example

Rebecca is 1.6 meters tall. Her brother Jamaal is 1.485 meters tall. How much taller is Rebecca than her brother?

Line up the numbers by their decimal points. Add two zeroes as placeholders at the end of 1.6. That way, both numbers have the same number of digits to the right of the decimal point. Then subtract.

```
    5910
  1.~~6~~~~0~~~~0~~
− 1.485
———————
  0.115
```

Rebecca is 0.115 meter taller than her brother.

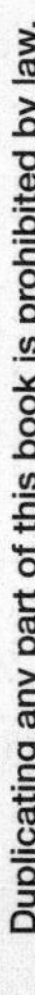

Practice

Directions: For questions 1 through 12, add.

1. $3.12 + $2.85 = ________

2. 18.08
   + 4.55

3. 55.8 + 8.311 = ________

4. 9.48
   + 5.27

5. 4.238 + 2.91 = ________

6. 23.4 + 18.09 = ________

7. 88.27
   + 55.74

8. 0.8 + 10.339 = ________

9. 3.707
   + 2.484

10. 6.65 + 4.4 = ________

11. Corrine earned $57.50 waiting tables, plus another $45.45 in tips. How much did she earn altogether?

    ____

12. Marco ran 2.67 miles to a county park. He then ran 1.5 miles in a loop around the park. How many miles did he run in total?

    ____

Directions: For questions 13 through 22, subtract.

13. 3.3 − 2.09 = ____________

14. 6.77
 − 3.49

15. $10.00 − $5.79 = ____________

16. 11.088
 − 5.505

17. 4.009 − 3.1 = ____________

18. 18.21 − 9.6 = ____________

19. 48.056
 − 29.048

20. 25 − 11.41 = ____________

21. 89.45
 − 77.54

22. 0.03 − 0.004 = ____________

23. A laptop computer weighs 6.4 pounds. A notebook computer weighs 3.75 pounds. How much heavier is the laptop than the notebook?

____

Explain how you found your answer.

____

____

____

Lesson 9: Multiplying and Dividing Decimals

Use the following steps to multiply decimals.

Step 1: **Multiply decimals as if they were whole numbers.**

Step 2: **Count the number of digits to the right of the decimal point in each factor.**

Step 3: **Move the decimal point that many places to the left in the product.**

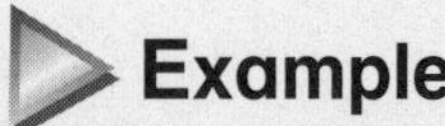

Example

Multiply: 4.5×245.76

Multiply decimals as if they were whole numbers.

$$\begin{array}{r} 245.76 \\ \times\ 4.5 \\ \hline 122880 \\ +\ 983040 \\ \hline 1{,}105{,}920 \end{array}$$

Count the number of digits to the right of the decimal point in each factor.

245.76 ← **2 digits to the right of the original decimal point**

× 4.5 ← **1 digit to the right of the original decimal point**

Move the decimal point that many places to the left in the product.

$$\begin{array}{r} 245.76 \\ \times\ 4.5 \\ \hline 122880 \\ +\ 983040 \\ \hline 1105.920 \end{array}$$

1105.920 ← **Move 3 digits to the left of the original decimal point.**

Therefore, $4.5 \times 245.76 = 1{,}105.92$.

CCSS: 6.NS.3

Use the following steps to divide decimals.

Step 1: **If the divisor is a decimal, move the decimal point to the right to make it a whole number. Move the decimal point in the dividend to the right the same number of places.**

Step 2: **Divide the decimals as if they were whole numbers.**

Step 3: **Move the decimal point from its new location in the dividend into the quotient.**

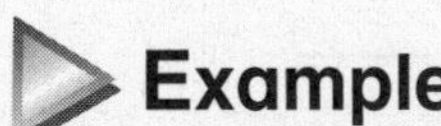

Example

Divide: 8.4 ÷ 2.4

Make the divisor a whole number by moving the decimal point to the right. Move the decimal point in the dividend to the right the same number of places.

```
2.4.)8.4.0
```

Divide the decimals as if they were whole numbers.

```
     3 5
24)84.0
  - 72
    12 0
  - 12 0
       0
```

Move the decimal point from its new location in the dividend into the quotient.

```
     3.5
24)84.0
  - 72
    12 0
  - 12 0
       0
```

Therefore, 8.4 ÷ 2.4 = 3.5.

Practice

Directions: For questions 1 through 14, find the product.

1. $5.1 \times 0.08 =$ ____________

2. $45.3 \times 2.34 =$ ____________

3. $3.7 \times 0.26 =$ ____________

4. $0.4 \times 0.9 =$ ____________

5. $12.98 \times 13 =$ ____________

6. $0.05 \times 1.2 =$ ____________

7. $6.09 \times 2.5 =$?

   A. 15.225
   B. 15.286
   C. 15.475
   D. 15.485

8. $5.2 \times 0.4 =$ ____________

9. $64.2 \times 3.4 =$ ____________

10. $7.91 \times 3.8 =$ ____________

11. $0.06 \times 8 =$ ____________

12. $0.62 \times 0.4 =$ ____________

13. $10.02 \times 0.1 =$ ____________

14. $13.2 \times 9.6 =$?

   A. 93.62
   B. 104.62
   C. 126.72
   D. 128.80

15. Frank ran the 100-meter dash in 10.9 seconds. It took Vern 1.25 times that long to run it. How long did it take Vern to run the 100-meter dash?

    ____

16. Cassidy earns $7.80 per hour for babysitting. How much would Cassidy earn if she babysat for 4.5 hours?

    ____

17. Malik bought 2.5 pounds of hamburger meat. The hamburger meat cost $3.18 per pound. How much did Malik pay for the hamburger meat?

    ____

18. Leah picked 2.6 pounds of blackberries. Ashlyn picked 1.8 times as many pounds of blackberries as Leah. How many pounds of blackberries did Ashlyn pick?

    ____

19. Corey bought 5 T-shirts for $9.79 each. How much did Corey spend on T-shirts?

    ____

20. Raj jogs an average of 0.63 of an hour each day. How much time does he spend jogging in 5 days?

    ____

21. Loretta swam 2 laps in 12.09 seconds. It took her 1.4 times as long to swim a third and fourth lap. How long did it take her to swim the third and fourth laps?

    ____

Directions: For questions 22 through 35, find the quotient.

22. 4.2 ÷ 3.5 = ____________

23. 1634 ÷ 1.9 = ____________

24. 10.62 ÷ 3 = ____________

25. 20.18 ÷ 0.2 = ____________

26. 195 ÷ 5.2 = ____________

27. 65 ÷ 1.3 = ____________

28. 0.26 ÷ 0.4 = ?

    A. 0.104
    B. 0.62
    C. 0.65
    D. 0.85

29. 0.9 ÷ 10 = ____________

30. 2.56 ÷ 1.6 = ____________

31. 0.56 ÷ 0.7 = ____________

32. 0.78 ÷ 5.2 = ____________

33. 3.6 ÷ 2.25 = ____________

34. 532 ÷ 2.8 = ____________

35. 66.24 ÷ 3.6 = ?

    A. 18.147
    B. 18.17
    C. 18.4
    D. 18.628

36. Lauren is helping her parents put a row of bricks in front of their garden. The length of the garden is 105 inches. If each brick is 7.5 inches long, how many bricks will be used for the row?

37. Strawberries are on sale for $2.30 per pound. Noah bought a bag of strawberries for $8.05. How many pounds of strawberries did Noah buy?

38. Alexis spent $12.72 on 8 equal-priced notebooks. How much did each notebook cost?

39. Justin made 6 banana shakes for his friends. He used a total of 7.5 bananas. How many bananas did Justin use in each shake?

40. Whitney's class went on a field trip to the St. Louis Gateway Arch. Each student rode the tram and saw a movie about the making of the arch. The total cost of the student tickets was $142.50. If each combined ticket for the tram and the movie cost $7.50, how many students went on the field trip?

41. Rachel used 1.2 gallons of paint to paint 225 square feet of wall. How much wall does one gallon of paint cover?

Explain how you found your answer.

Lesson 10: Dividing Fractions

An area model can help show how to divide fractions less than 1.

Example

A farmer plants cucumbers on $\frac{1}{2}$ acre of land. She would need $\frac{2}{3}$ acre to supply her entire town with cucumbers. How much of her town's demand for cucumbers can the farmer supply?

To solve this problem, create an area model to show $\frac{1}{2}$ of an acre. Then create rows to show the fraction of an acre the town needs, $\frac{2}{3}$.

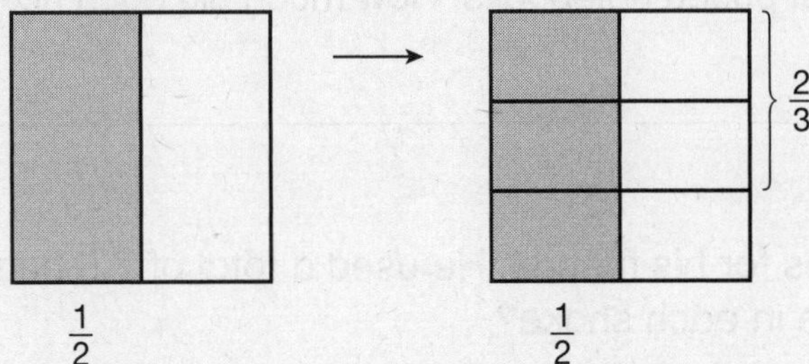

The denominator of the quotient is the number of squares in the rows that show $\frac{2}{3}$. It may help to move any shaded squares into the rows for $\frac{2}{3}$.

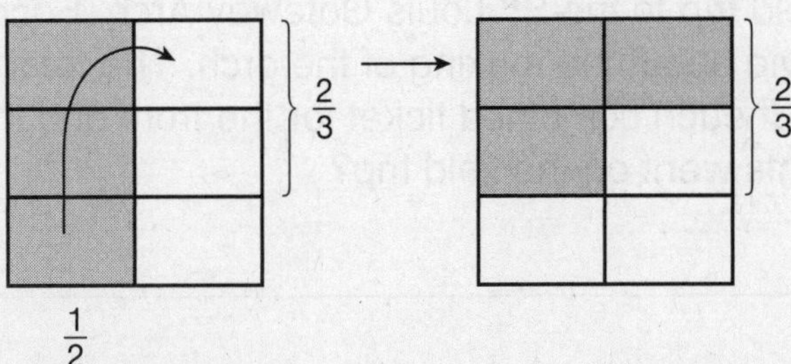

There are 4 squares in the rows that show $\frac{2}{3}$. These squares represent the parts of an acre that should be planted with cucumbers to supply the town.

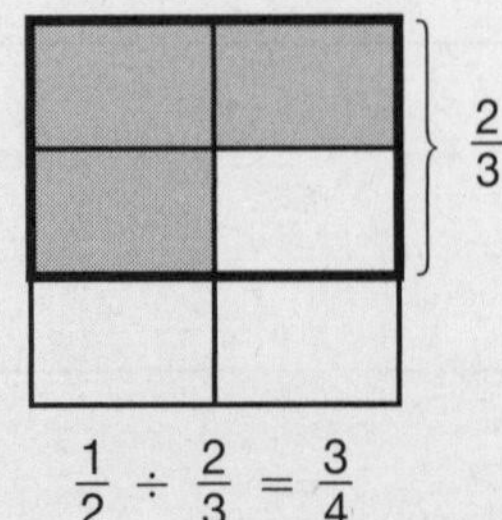

The farmer can supply $\frac{3}{4}$ of her town's demand for cucumbers.

You can also divide fractions by multiplying the dividend by the **reciprocal** of the divisor. To find the reciprocal of a fraction, switch its numerator and denominator. Then multiply the two fractions by multiplying the numerators and the denominators.

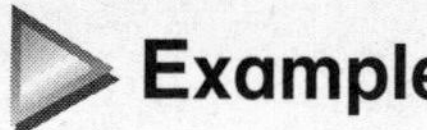

Example

Divide: $\frac{5}{8} \div \frac{3}{4}$

Change the divisor to its reciprocal, and change the division symbol to a multiplication symbol.

$\frac{5}{8} \div \frac{3}{4}$ becomes $\frac{5}{8} \times \frac{4}{3}$

Multiply the numerators and denominators. Write the answer in simplest form.

$$\frac{5 \times 4}{8 \times 3} = \frac{20}{24} = \frac{5}{6}$$

Therefore, $\frac{5}{8} \div \frac{3}{4} = \frac{5}{6}$.

To divide mixed numbers, first change them into improper fractions. Then divide them as you would proper fractions.

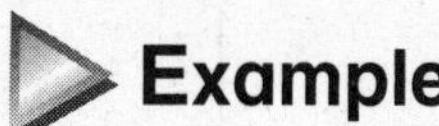

Example

Divide: $2\frac{1}{4} \div \frac{7}{9}$

Change the mixed number to an improper fraction.

$$2\frac{1}{4} = \frac{2 \times 4 + 1}{4} = \frac{9}{4}$$

Change the divisor to its reciprocal. Then change the division symbol to a multiplication symbol.

$\frac{9}{4} \div \frac{7}{9}$ becomes $\frac{9}{4} \times \frac{9}{7}$

Multiply the numerators and denominators. Write the answer in simplest form.

$$\frac{9 \times 9}{4 \times 7} = \frac{81}{28} = 2\frac{25}{28}$$

Therefore, $2\frac{1}{4} \div \frac{7}{9} = 2\frac{25}{28}$.

TIP: If you are dividing with a whole number, remember to change it to an improper fraction. For example, $5 = \frac{5}{1}$.

Practice

Directions: For questions 1 through 5, divide the fractions. Use the blank squares as area models.

1. $\frac{1}{2} \div \frac{3}{4} =$ ____________

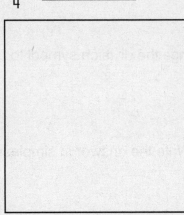

2. $\frac{2}{3} \div \frac{3}{4} =$ ____________

3. $\frac{1}{4} \div \frac{4}{5} =$ ____________

4. $\frac{2}{5} \div \frac{1}{2} =$ ____________

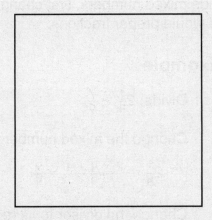

5. Michaela made $\frac{1}{3}$ pound of cookie batter. She needs $\frac{1}{2}$ pound to make a full batch of cookies. What fraction of a full batch of cookies can Michaela make?

____

Directions: For questions 6 through 13, divide the fractions. Leave your answers in simplest form.

6. $\frac{9}{10} \div \frac{2}{3} =$ ____________

7. $\frac{1}{8} \div \frac{3}{4} =$ ____________

8. $\frac{2}{5} \div \frac{3}{5} =$ ____________

9. $\frac{4}{7} \div \frac{1}{2} =$ ____________

10. $\frac{7}{10} \div \frac{4}{9} =$ ____________

11. $\frac{1}{5} \div \frac{2}{15} =$ ____________

12. $\frac{7}{12} \div \frac{4}{5} =$ ____________

13. $\frac{7}{8} \div \frac{1}{3} =$ ____________

Directions: For questions 14 through 23, divide the fractions and mixed numbers. Leave your answers in simplest form.

14. $1\frac{1}{2} \div \frac{2}{5} =$ ____________

15. $\frac{1}{3} \div 1\frac{1}{3} =$ ____________

16. $2\frac{1}{5} \div 1\frac{1}{5} =$ ____________

17. $3\frac{1}{4} \div \frac{3}{8} =$ ____________

18. $1\frac{9}{10} \div \frac{2}{5} =$ ____________

19. $2\frac{3}{8} \div \frac{3}{4} =$ ____________

20. $\frac{11}{12} \div 1\frac{1}{6} =$ ____________

21. $1\frac{2}{3} \div \frac{1}{4} =$ ____________

22. $3\frac{3}{4} \div 2\frac{1}{2} =$ ____________

23. $\frac{1}{8} \div 1\frac{1}{4} =$ ____________

24. Howard bought $\frac{3}{5}$ pound of granola. He divided the granola evenly among 4 friends. How much granola did each friend get?

____

25. Jasmine has $\frac{1}{4}$ of a pizza left. She is going to divide the leftovers into 6 equal pieces. What fraction of the whole pizza will each piece be?

____

26. Yolanda has $4\frac{1}{4}$ cups of flour. She needs $1\frac{2}{3}$ cups of flour to make a batch of cookies. How many batches of cookies can Yolanda make with the flour she has?

____

27. A businesswoman uses $\frac{1}{3}$ of the floor of an office building. She needs to use $\frac{5}{8}$ of the floor to expand her business. How much of the space that she needs does she currently use?

____

28. Maximilian had $\frac{4}{5}$ bag of trail mix. He divided the trail mix equally among 5 friends. What fraction of the full bag of trail mix did each friend receive?

____

29. Jeremiah bought a plank of wood $2\frac{2}{3}$ yards long. He divided the plank into 3 equal pieces. How long is each of Jeremiah's planks?

____

Explain how you found your answer.

____

____

____

Unit 1 Practice Test

Read each question. Choose the correct answer.

1. A hardware store has 695 nails in a bin. The store owner wants to put them into boxes with 15 nails in each box. How many boxes can the owner fill, and how many will be left over?

   A. 45 with 10 left over

   B. 46 with none left over

   C. 46 with 3 left over

   D. 46 with 5 left over

2. After a birthday party, $\frac{2}{5}$ of a cake is left over. The remaining cake is split into 4 equal pieces. What fraction of the original cake is each piece?

   A. $\frac{1}{20}$

   B. $\frac{1}{10}$

   C. $\frac{2}{9}$

   D. $\frac{8}{5}$

3. A grocery store charges $1.78 per pound of seedless grapes. How much does the store charge for 3.5 pounds of seedless grapes?

   A. $5.23

   B. $6.23

   C. $14.24

   D. $62.30

4. Which is the value of $|100|$?

   A. -100

   B. $\frac{1}{100}$

   C. 10

   D. 100

5. What is the greatest common factor of 9 and 15?

   A. 3

   B. 6

   C. 45

   D. 135

6. The number of yards carried by four football players during a game is shown in the following table.

Football Player	Yards Carried
Antoine	32
Robbie	−5
Stephen	28
Quentin	−1

Which shows the order of the football players from **fewest** yards carried to **most** yards carried?

A. Quentin, Robbie, Stephen, Antoine

B. Robbie, Quentin, Stephen, Antoine

C. Robbie, Quentin, Antoine, Stephen

D. Antoine, Stephen, Quentin, Robbie

7. A scuba diver dived 18 meters below the surface. Which represents the total distance that the diver dived?

A. −18 meters

B. 0 meters

C. 18 meters

D. 81 meters

8. What is the least common multiple of 2 and 10?

____

9. Maude uses $\frac{3}{4}$ of her allowance to get a massage for $\frac{3}{5}$ of an hour. What fraction of her allowance would she need to get a massage for a full hour? Use the following area model to help solve.

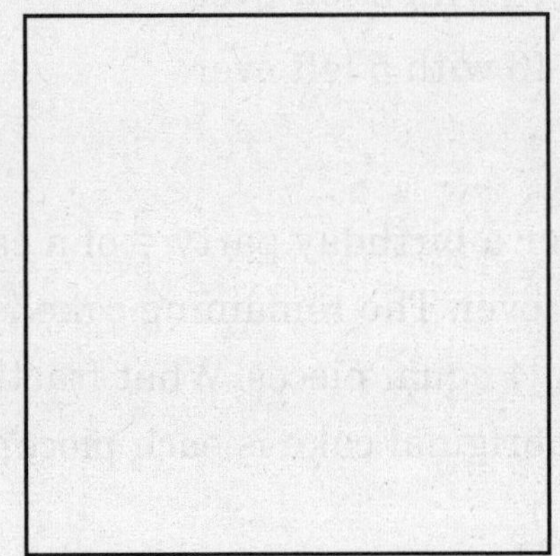

____

10. Points A, B, C, D, and E are plotted in the following coordinate plane.

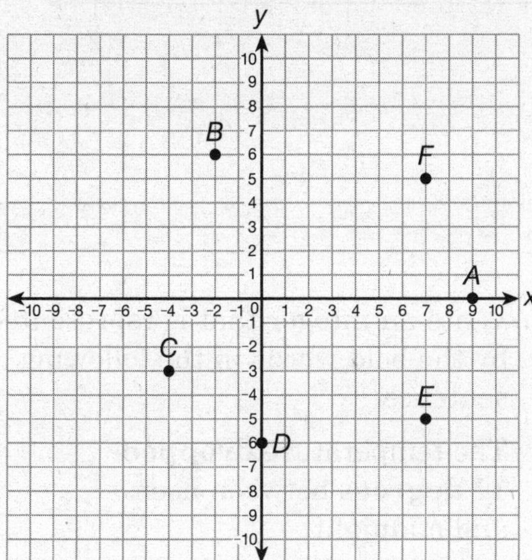

What point is represented by the ordered pair (7, −5)?

____

11. Write an integer that is represented by the bold words in the following sentence.

Jonathan spent **20 minutes less** on his science homework than he did on his math homework.

____

12. Taylor buys a keyboard for $32.78. He pays for it with a $50 bill. How much change should Taylor get back?

____

13. Of the ten largest American cities, the four that experienced the smallest change in population from 1990 to 2000 were Detroit, Chicago, San Diego, and Philadelphia. The following table shows the change in population from 1990 to 2000 for those cities.

City	Change in Population (1990–2000)
Detroit	−76,704
Chicago	112,290
San Diego	112,851
Philadelphia	−68,027

Write the American cities in order from the **greatest** population change to **least** population change.

____

14. Point A on a coordinate plane has coordinates of (4, −3). The point is reflected over both the x- and the y-axes. What are the coordinates of the new point?

____

15. Eliza uses a number line to compare $\frac{3}{10}$ and $-\frac{7}{10}$.

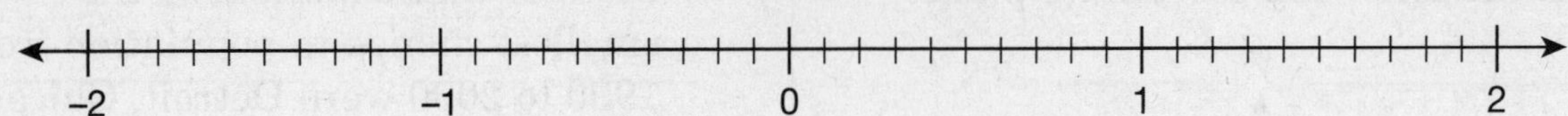

Compare Eliza's numbers using <, >, or =.

16. $5\frac{1}{2} \div \frac{1}{3} =$

17. What is the GCF of 92 and 100?

18. $11\overline{)878}$ =

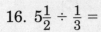

19. Write an integer that is represented by the bold words in the following sentence.

The temperature **dropped 17 degrees** between noon and midnight.

20. What is the value of $|0|$?

21. What is the opposite of -25?

22. After pulling into a parking spot, Ms. Feliz backed her car up 2 feet. Write an expression with absolute value to show the distance Ms. Feliz traveled after pulling into the parking spot.

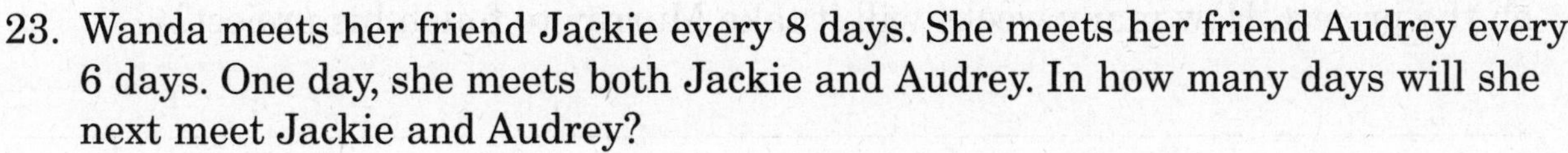

23. Wanda meets her friend Jackie every 8 days. She meets her friend Audrey every 6 days. One day, she meets both Jackie and Audrey. In how many days will she next meet Jackie and Audrey?

24. The following rectangular park has an area of $\frac{1}{8}$ km.

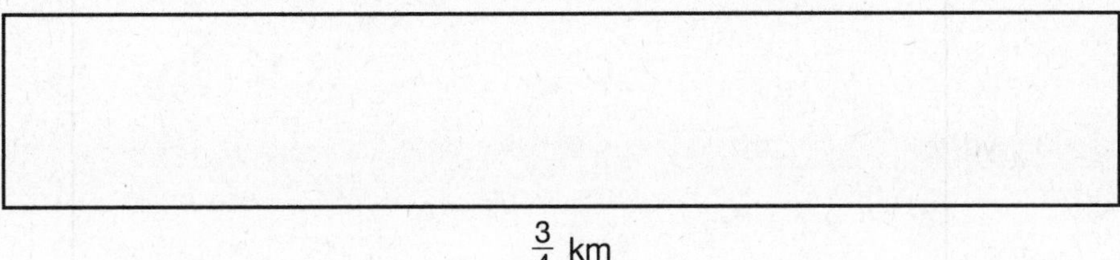

$\frac{3}{4}$ km

What is the width of the park?

25. Find the difference.

$$\begin{array}{r} 13.58 \\ -\ 9.62 \\ \hline \end{array}$$

26. $998 \div 32 =$ __________

27. What is the GCF of 1 and 10?

28. Isabella collected $\frac{1}{4}$ gallon of honey from her beehive. She wants to fill her jar, which holds $\frac{3}{10}$ gallon. What fraction of her jar has Isabella already filled with honey? Use the following model to help find the answer.

29. Find the product.

$$\begin{array}{r} 8.84 \\ \times\ 5.25 \\ \hline \end{array}$$

30. Murray needs to spend 700 hours on a project. He can spend 20 hours each week on the project. How many weeks will it take Murray to finish his project?

____

31. Mitchell is hired to paint a mural on a wall with an area of $21\frac{1}{8}$ square yards.

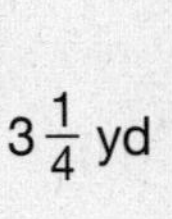

What is the length of the wall?

____

32. Write the following statement using <, >, or =.

5 yards is less than 8 yards.

____

Explain how you were able to interpret the statement using symbols.

____

____

____

33. The following menu shows the prices for snacks at Farah's Food Shack.

Farah's Food Shack Menu

Snacks		Drinks	
Hot dog:	$2.45	Juice:	$1.73
Pretzel:	$2.25	Lemonade:	$1.35
Granola bar:	$0.89	Water:	$1.15

Part A

A group of hikers orders 8 granola bars at Farah's Food Shack. What is the cost of those snacks?

____

Part B

The hikers add 1 juice and 1 water to their order. What is the total price of their order in addition to the snacks?

____

Part C

A sales tax rate of 0.06 is charged on the snacks and drinks. How much extra will the hikers have to pay for the sales tax?

____

Explain how you found the total amount of sales tax that the hikers had to pay.

____

____

____

____

34. Point D is shown on the following coordinate plane.

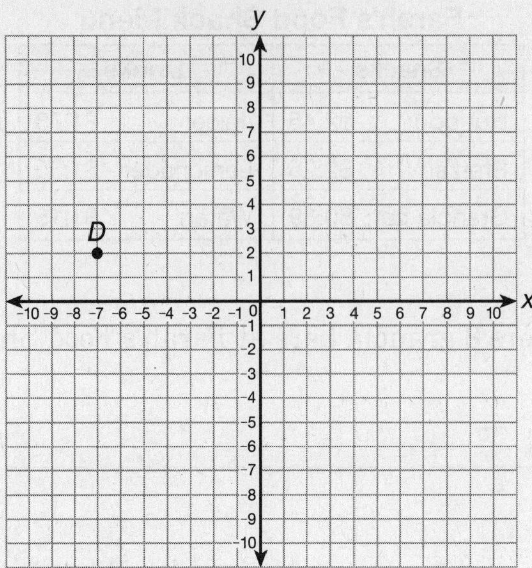

Part A

What are the coordinates of point D?

____

Part B

Point E has the same coordinates as point D except that the y-coordinate is opposite in sign. Plot point E in the coordinate plane above.

Part C

What are the coordinates of point E?

____

Explain how you knew the location of point E.

____

____

____

____

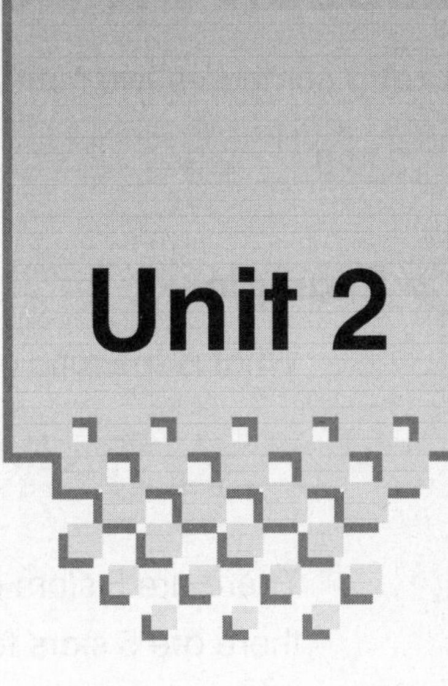

Ratios and Proportional Relationships

An airplane can fly from New York City to Miami, a distance of about 1000 miles, in two hours. Using this information, you can determine the speed of the plane in miles per hour. You can also determine how long it would take the plane to fly any distance, such as 2000 miles from New York City to Salt Lake City. To find this information, you need to use unit rates and ratios.

In this unit, you will learn about the different ways that ratios can be represented. You will learn to find the percent of a number and find a number given the percent and a part. You will identify the unit rate in a situation, and you will solve problems using those unit rates. You will also solve problems involving a coordinate plane, including plotting points to represent a rate or a ratio. Finally, you will use ratios to convert measurements in length, weight, and capacity.

In This Unit

Lesson 11: Ratios

A **ratio** compares two numbers. Ratios can be written in the following ways.

2 to 3 2:3 $\frac{2}{3}$

Example

What is the ratio of stars to moons? What is the ratio of moons to stars?

There are 5 stars and 3 moons. The ratio of stars to moons is 5 to 3, since there are 5 stars for every 3 moons. This is written as $\frac{5}{3}$ or 5:3.

The ratio of moons to stars is 3 to 5, because there are 3 moons for every 5 stars. This is written as $\frac{3}{5}$ or 3:5.

Example

What is the ratio of pens to erasers?

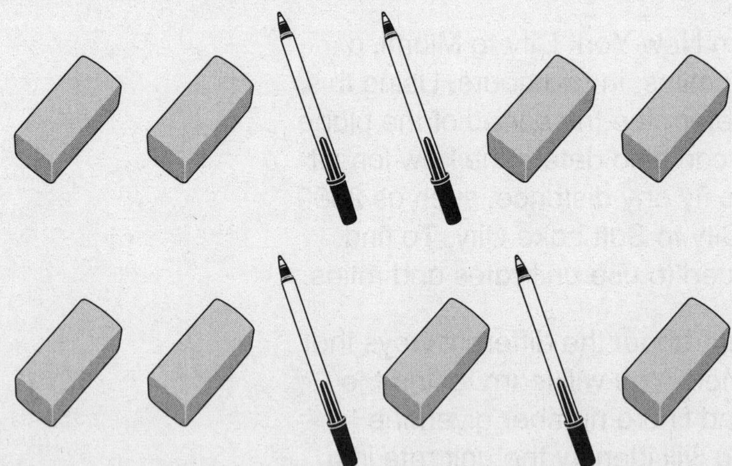

There are 4 pens and 8 erasers. The ratio of pens to erasers can be written as 4 to 8, 4:8, or $\frac{4}{8}$. The ratio can also be written in simplest form as 1 to 2, 1:2, or $\frac{1}{2}$.

CCSS: 6.RP.1

Ratios can also be expressed with words. Ratios can be used to describe the relationship between two quantities.

Example

Zander has paper clips and pushpins in his desk drawer. Describe the relationship between paper clips and pushpins.

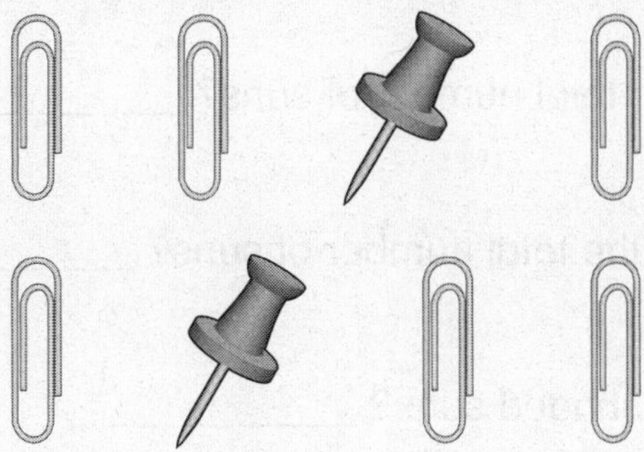

There are 6 paper clips and 2 pushpins. The ratio of paper clips to pushpins in Zander's drawer is, therefore, 6:2, or 3:1. This means that for every 3 paper clips, there is 1 pushpin.

You can also use ratios to compare the items to a whole. For every 4 items in Zander's desk drawer, there are 3 paper clips and 1 pushpin. You can write the ratio of items in the desk drawer to paper clips as 4 to 3, 4:3, or $\frac{4}{3}$. The ratio of items in the desk drawer to pushpins is 4 to 1, 4:1, or $\frac{4}{1}$.

Example

The following table shows the results of a small town's election for mayor. Describe the relationship between the candidates' votes using ratio language.

Candidate	Number of Votes
Natalie Mallot	300
James Porter	200

Natalie Mallot received 300 votes compared to 200 votes for James Porter. The ratio of votes for Mallot to votes for Porter is 300:200, or 3:2. For every 3 votes Mallot received, Porter received 2 votes. For every 5 votes in the election, Mallot received 3 votes and Porter received 2 votes.

Practice

Directions: Use the following drawing to answer questions 1 through 3.

1. What is the ratio of shaded suns to the total number of suns? ____________

2. What is the ratio of unshaded suns to the total number of suns? ____________

3. What is the ratio of shaded suns to unshaded suns? ____________

Directions: Use the following drawing to answer questions 4 through 6. Write your answers as fractions in simplest form.

4. What is the ratio of faces to hearts? ____________

5. What is the ratio of hearts to faces? ____________

6. What is the ratio of hearts to the total number of faces and hearts? ____________

7. What is the ratio of shaded pens to unshaded pens in simplest form? ____________

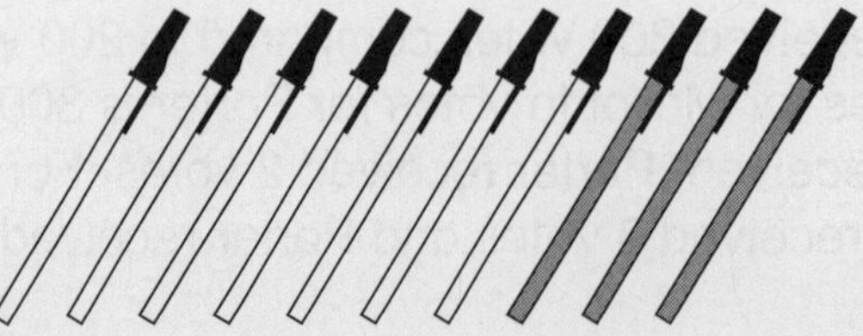

CCSS: 6.RP.1

Directions: Use the following drawing to answer questions 8 through 10.

8. Describe the relationship of the number of cats to the number of dogs.

___

9. Describe the relationship of the number of dogs to the number of cats.

___

10. Describe the relationship of the number of cats to the number of animals in total.

___

11. The drawing shows a package of eight eggs. Using ratios, describe the relationship of the number of brown eggs to white eggs. Describe the ratio in simplest form.

    ___

    Explain how you found your answer, including whether you were able to simplify or not.

    ___

    ___

Lesson 12: Percents

The word *percent* is made up of two parts, *per* meaning "out of", and *cent* meaning "one hundred". *Percent* literally means "out of each hundred." A **percent** represents the parts of a whole that is divided into 100 equal parts. The following grid represents a whole unit, or 100%. The grid shows that 16 squares out of 100 are shaded. Therefore, 16% of the grid is shaded.

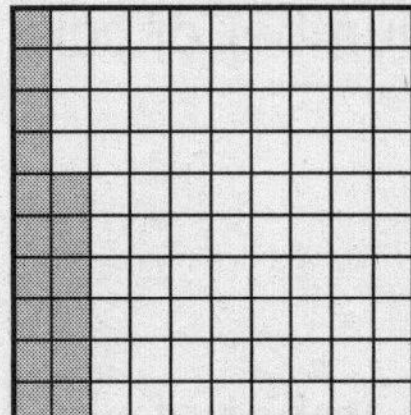

You can consider a percent as a ratio per 100. In the previous grid, there are 16 shaded squares. There are 100 total squares. The ratio of shaded squares to total squares is 16:100, or $\frac{16}{100}$. If the second part, or denominator, of a ratio is 100, then the first part, or numerator, is the percent.

Example

The ratio of horses to all the animals on a farm is $\frac{11}{100}$. What percent of the animals on the farm are horses?

Because the denominator of the ratio is 100, the numerator must be the percent. Of all the animals on the farm, 11% are horses.

Example

Luz got 88% of the questions right on her science quiz. What ratio represents the number of questions she got right to the total number of questions?

To convert the percent into a ratio, you can use the percent as the numerator and add a denominator of 100. The ratio of questions Luz got right to the total number of questions is, therefore, $\frac{88}{100}$, which can be simplified to $\frac{22}{25}$. The ratio can also be shown as 22:25 or 22 to 25.

TIP: A ratio should be simplified using common factors. However, a percent should always be out of 100.

CCSS: 6.RP.3.c

If you know a percent and the whole, you can find the part of the whole that the percent represents. You first need to convert the percent to a ratio as a fraction. Then multiply the whole by the fraction. The product will be that part of the whole.

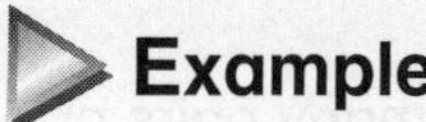

Example

What is 60% of 30?

A tape diagram can help show the relationship of the whole and the part.

30
60% of 30

The percent 60% is equal to the ratio $\frac{60}{100}$. Multiply the whole by this ratio. You will need to convert 30 to $\frac{30}{1}$.

$$\frac{30}{1} \times \frac{60}{100} = \frac{30 \times 60}{1 \times 100} = \frac{1800}{100} = 18$$

60% of 30 is 18.

If you know a percent and the part of the whole, you can find the total number of items in the whole. You first need to convert the percent to a ratio as a fraction. Then you can divide the part by the fraction. The quotient will be the total number.

Example

85% of what number is equal to 68?

The tape diagram shows the relationship of the whole and the part.

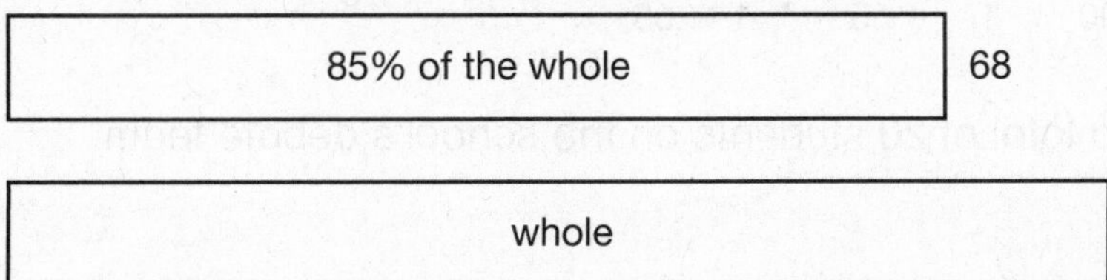

The percent 85% is equal to the ratio $\frac{85}{100}$. Divide the part by this ratio.

$$\frac{68}{1} \div \frac{85}{100} = \frac{68}{1} \times \frac{100}{85} = \frac{68 \times 100}{1 \times 85} = \frac{6800}{85} = 80$$

85% of 80 is equal to 68.

You may need to solve for the part or the whole with a percent when working with real-world problems.

Example

Kevin has 15 pairs of socks. Of his socks, 40% are black. How many pairs of black socks does Kevin have?

This problem provides the total number of pairs of socks and the percent that are black. You need to find the part of the total number. That part will be the number of black socks.

The first step is to convert the percent to a ratio as a fraction: $40\% = \frac{40}{100}$. Then multiply the total by this ratio to find the part. You will need to convert 15 to $\frac{15}{1}$.

$$\frac{15}{1} \times \frac{40}{100} = \frac{15 \times 40}{1 \times 100} = \frac{600}{100} = 6$$

Kevin has 6 pairs of black socks.

Example

There are 13 girls on a school's debate team. If 65% of the team members are girls, how many students are on the debate team in total?

This problem provides the number of girls on a team and the percent that are girls. You need to find the total number. That number will represent the total number of students on the debate team.

The first step is to convert the percent to a ratio as a fraction: $65\% = \frac{65}{100}$. Then divide the part by this ratio to find the whole. You will need to convert 13 to $\frac{13}{1}$.

$$\frac{13}{1} \div \frac{65}{100} = \frac{13}{1} \times \frac{100}{65} = \frac{13 \times 100}{1 \times 65} = \frac{1300}{65} = 20$$

There are a total of 20 students on the school's debate team.

CCSS: 6.RP.3.c

Practice

Directions: For questions 1 through 3, find the percent of the grid that is shaded.

1. 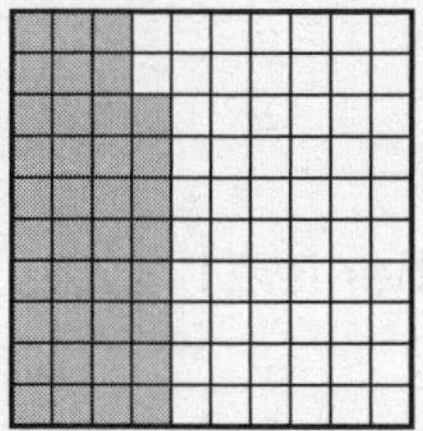 ____________

2. 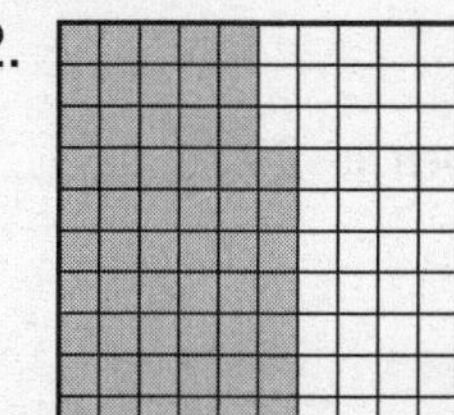 ____________

3. 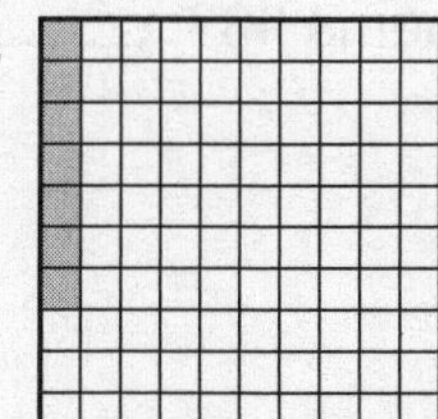 ____________

Directions: For questions 4 through 6, shade in the given percent of each grid.

4. 53% 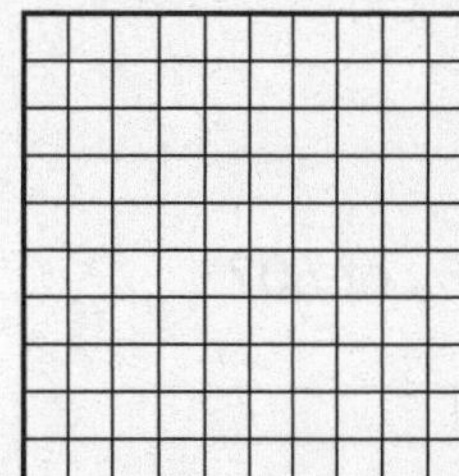

5. 88% 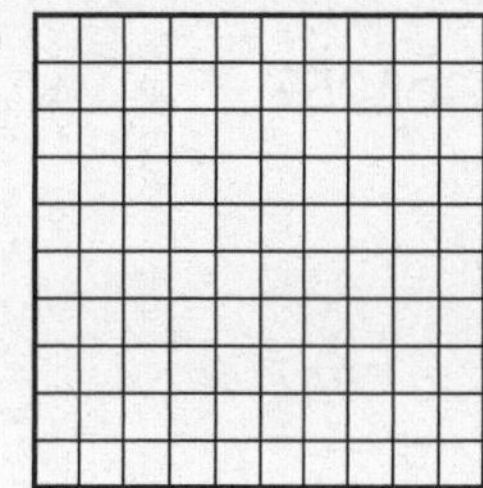

6. 17% 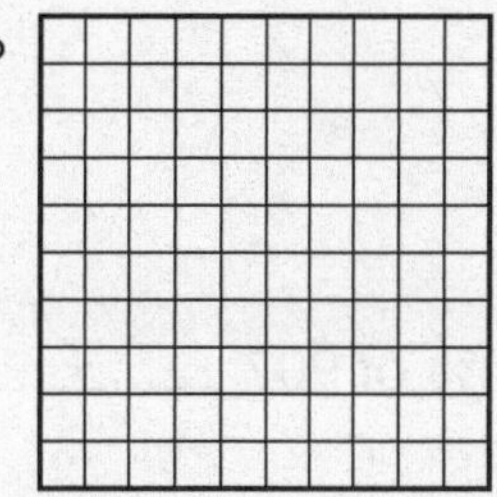

7. The ratio of teenagers to all the customers in a mall is $\frac{36}{100}$. What percent of the customers in a mall are teenagers?

____

8. Of all the students in Ms. Mack's class, 44% are girls. What ratio compares the number of girls in her class to the total number of students in her class?

____

Directions: For questions 9 through 22, find the part or the whole of a percent.

9. What is 32% of 50? __________

10. What is 15% of 60? __________

11. What is 8% of 25? __________

12. What is 75% of 92? __________

13. What is 80% of 90? __________

14. What is 25% of 88? __________

15. What is 10% of 30? __________

16. 2% of what number is 1? __________

17. 30% of what number is 21? __________

18. 50% of what number is 29? __________

19. 75% of what number is 48? __________

20. 7% of what number is 7? __________

21. 25% of what number is 13? __________

22. 40% of what number is 26? __________

CCSS: 6.RP.3.c

23. It rained during 40% of the days of Chrissy's vacation. If it rained on 2 days, how many days long was Chrissy's vacation?

____

24. Only 20% of the children at a party voted to have tacos for lunch. There were 20 children at the party. How many voted to have tacos?

____

25. Mr. Malkmus replaced 75% of the wooden boards on his deck. If he replaced 63 wooden boards, how many boards does his deck have in total?

____

26. Of the families in Mel's apartment building, 55% have at least one pet. There are 80 families in Mel's apartment building. How many have at least one pet?

____

27. Raymond answered 95% of the questions on his English quiz correctly. If he answered 19 questions correctly, how many questions were there on the quiz in total?

____

28. Gillian has completed 25% of a bicycle race. If the length of the race is 32 km, how many km has she traveled so far?

____

29. Alain has spent 40 minutes on his homework. If he is 40% done, how many minutes will he spend on his homework in total?

____

Explain your answer.

____

____

____

Lesson 13: Rates and Unit Rates

A **rate** is a ratio that compares two different units. Some examples of rates are 72 miles per 3 gallons, $6 for 3 pounds, and 130 heartbeats in 2 minutes.

A **unit rate** is a rate that compares two different units, where one of the measurements is 1. Some unit rates are 24 miles per gallon, $2 per pound, and 65 heartbeats per minute. Notice that these unit rates could also be stated as 24 miles per 1 gallon, $2.00 per 1 pound, and 65 heartbeats per 1 minute.

Example

Mr. Easton bought 12 gallons of gas for $36. What is Mr. Easton's unit rate for the cost of 1 gallon of gas?

Use the ratio of dollars to gallons of gas to find the unit rate.

$$\frac{\text{dollars}}{\text{gallons of gas}} = \frac{36}{12}$$

$$= \frac{3}{1}$$

$$= 3$$

Mr. Easton paid $3 for every gallon of gas.

Example

Julia traveled 432 miles in 8 hours. What is the unit rate for the average number of miles she traveled per 1 hour?

Use the ratio of miles to hours to find the unit rate.

$$\frac{\text{miles}}{\text{hours}} = \frac{432}{8}$$

$$= \frac{54}{1}$$

$$= 54$$

Julia traveled an average of 54 miles per 1 hour.

CCSS: 6.RP.2, 6.RP.3.b

A tape diagram can help solve a rate problem. You can use it to find the unit rate. Then you can multiply the unit rate by the number of items.

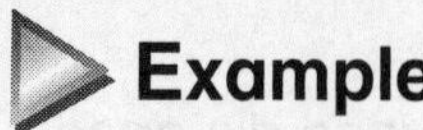

Example

The cost of 4 curtains is $48. What is the cost of 6 curtains?

The following tape diagram shows $48 split into 4 equal parts. The value of each part represents the cost of 1 curtain.

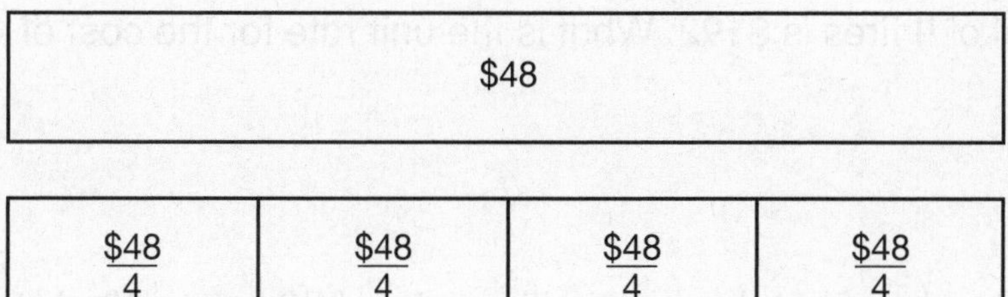

$\$48 \div 4 = \$12$

The unit rate is $12 for 1 curtain, so the cost of 6 curtains will be equal to the unit rate times 6.

$\$12 \times 6 = \$72$

The cost of 6 curtains is $72.

Example

The cost of 5 toy cars is $35. What is the cost of 9 toy cars?

The following tape diagram shows $35 split into 5 equal parts. The value of each part represents the cost of 1 toy car.

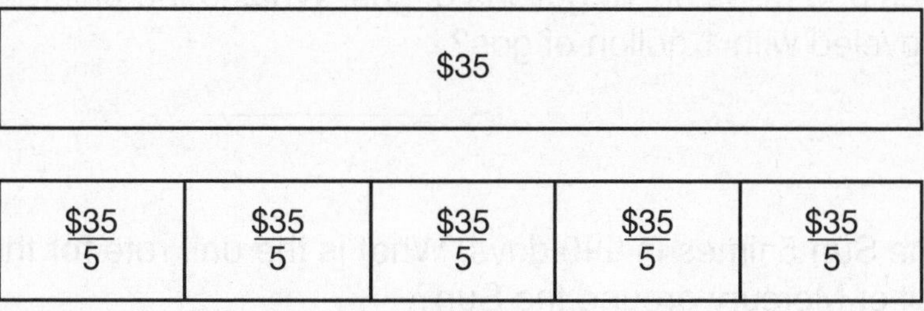

$\$35 \div 5 = \$7$

The unit rate is $7 for 1 toy car, so the cost of 9 toy cars will be equal to the unit rate times 9.

$\$7 \times 9 = \$63$

The cost of 9 toy cars is $63.

Practice

Directions: For questions 1 through 7, find the unit rate.

1. Xavier runs 21 miles in 3 hours. What is Xavier's unit rate for the average distance he ran after 1 hour?

   ____

2. The cost of a set of 4 tires is $192. What is the unit rate for the cost of 1 tire?

   ____

3. A chef bought 8 packs of hot dog buns with a total of 48 buns. What is the unit rate for the number of buns in 1 pack?

   ____

4. A physical trainer measures her heart rate and counts that it beats 30 times in 15 seconds. What is the unit rate for her heartbeat in 1 second?

   ____

5. A block of silver weighing 8 ounces is worth $104. What is the unit rate for the value of the silver in 1 ounce?

   ____

6. Ms. Minkins drives 338 miles on 13 gallons of gas. What is the unit rate for the average miles traveled with 1 gallon of gas?

   ____

7. Mercury orbits the Sun 5 times in 440 days. What is the unit rate for the number of days for 1 orbit of Mercury around the Sun?

   ____

Directions: For questions 8 through 13, solve the rate problem by using the unit rate. You may draw a tape diagram.

8. In a 3-minute period, 33 waves crash on the shore. How many waves will crash on the shore in 10 minutes?

 ____

9. A construction crew builds 2 floors of a skyscraper every 10 days. How long will it take for the crew to build 15 floors on the skyscraper?

 ____

10. A yogurt company charges 72 cents for a 6-ounce container of yogurt. How much does it charge for an 8-ounce container?

 ____

11. Ms. Goodwin earns $2570 for each 2-week pay period. How much will she make in 28 weeks?

 ____

12. Yuri needed 24 minutes to solve 6 math problems. How long will he need to solve 10 math problems?

 ____

13. A taxi charges $14 for a 7-mile trip. How much will the taxi charge for a 16-mile trip?

 ____

14. Ivana's printer prints 40 pages in 5 minutes. How many pages will the printer print in 8 minutes?

 ____

 Explain how you found your answer.

 ____

 ____

 ____

Lesson 14: Solving Problems Using Graphs

The graph of a rate or ratio can help show patterns. To graph a rate or ratio, first create a table of values. Then plot the values as ordered pairs.

Example

For every hour that Wanda drives on the highway, she uses 2 gallons of gasoline. Represent Wanda's rate of gasoline usage with a graph on a coordinate plane.

Before you can graph the rate, create a table of values. Multiply the rate by whole numbers. For instance, if Wanda drives $1 \times 3 = 3$ hours, she uses $2 \times 3 = 6$ gallons of gasoline.

Hours Driven on Highway, *x*	Gallons of Gasoline Used, *y*
1	2
2	4
3	6
4	8

Now you can represent each pair of values from the rows in the table as an ordered pair. Use the hours driven on the highway for *x* and the gallons of gasoline used for *y*. The following are ordered pairs: (1, 2), (2, 4), (3, 6), (4, 8).

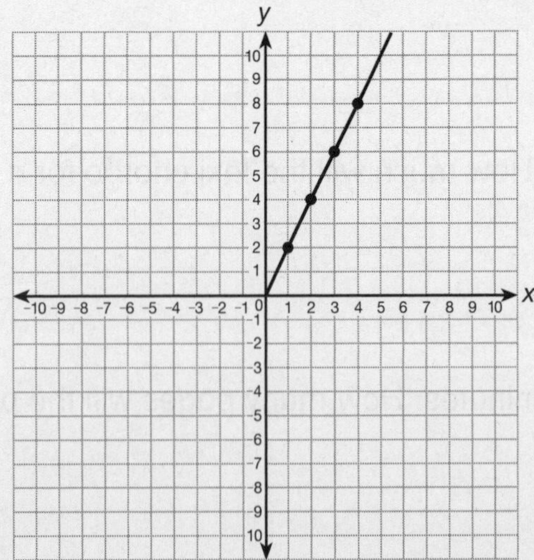

The line connecting the ordered pairs helps show the relationship between the two variables: the hours driven and the gallons of gasoline used. You can continue to plot additional ordered pairs onto the grid. For example, if Wendy drives 5 hours, she will use 10 gallons of gasoline. You can plot (5, 10) on the graph.

CCSS: 6.RP.3.a, 6.EE.9

You can compare more than one rate or ratio on the same graph.

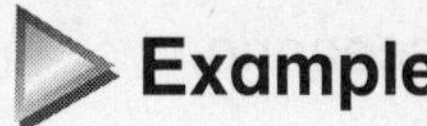

Example

For every 2 minutes, Desmond reads 3 pages in his book. Esmeralda created the table below, recording the numbers of pages she read in so many minutes. Graph the readers' rates to compare them.

Esmeralda	
Minutes	**Pages Read**
3	5
6	10
9	
	20

To graph their reading rates, create a table for Desmond's reading rate and complete Esmeralda's table. To create Desmond's table, multiply his rate by whole numbers. For instance, if Desmond reads for $2 \times 2 = 4$ minutes, he will read $3 \times 2 = 6$ pages.

Esmeralda reads 5 pages for every 3 minutes. If Esmeralda reads for $3 \times 3 = 9$ minutes, she will read $5 \times 3 = 15$ pages. If Esmeralda reads $5 \times 4 = 20$ pages, she has read for $3 \times 4 = 12$ minutes.

Desmond	
Minutes	**Pages Read**
2	3
4	6
6	9
8	12

Esmeralda	
Minutes	**Pages Read**
3	5
6	10
9	**15**
12	20

Represent each pair of values from the table as an ordered pair. Use the minutes for *x* and the pages read for *y*. Because it is not possible to read for a negative number of minutes, the graph shows the first quadrant only.

The steepness of the lines shows how quickly Desmond and Esmeralda read. Because the line representing Esmeralda's reading rate is steeper, it means that she reads more pages per minute. Esmeralda reads at a faster rate.

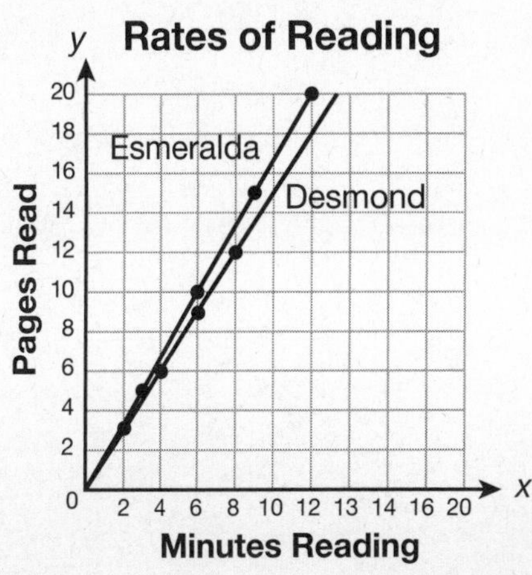

Practice

Directions: For questions 1 and 2, create a table of values for the given scenario. Then graph the points on the coordinate plane.

1. During the first 6 months of Audrey's life, she grew 1 inch each month.

Audrey's Age (in months)	Growth (in inches)

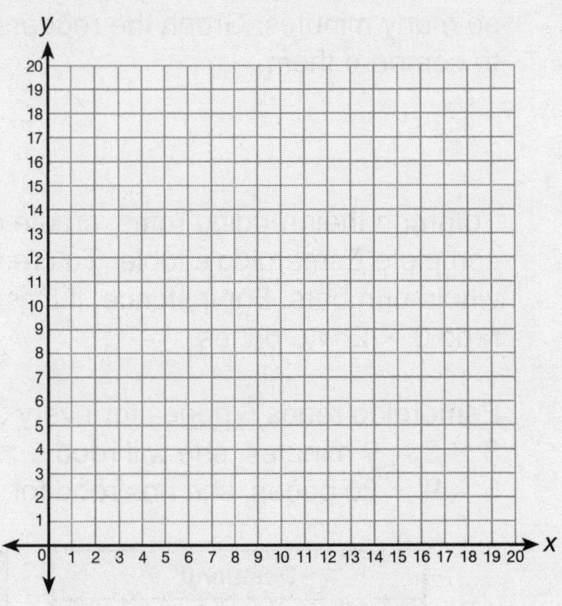

2. Roberto times his workouts at the gym. Every 4 seconds, he can do 3 push-ups.

Number of Seconds	Number of Push-ups

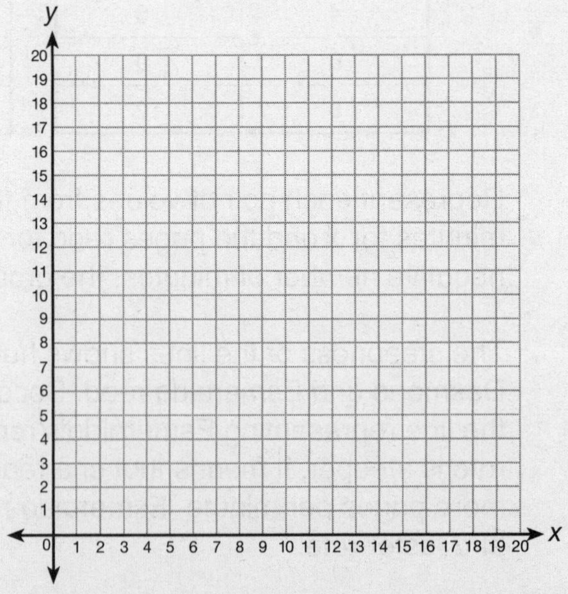

CCSS: 6.RP.3.a, 6.EE.9

Directions: For questions 3 and 4, create tables of values for the two rates or ratios in each scenario. Then graph the points on the coordinate plane to compare the rates or ratios.

3. Isabella walks at a rate of 3 miles per hour. Jordan walks at a rate of 4 miles per hour.

Isabella	
Hours	**Miles Walked**

Jordan	
Hours	**Miles Walked**

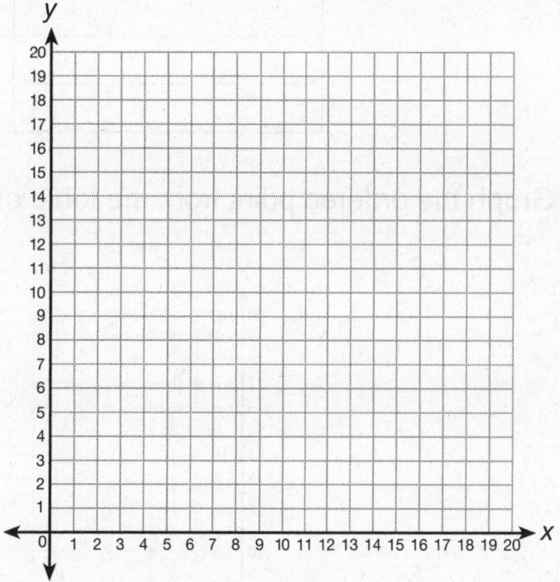

4. For each 4 free throws taken, Jasmine makes 3 baskets. For each 3 free throws taken, Ivan makes 2 baskets.

Jasmine	
Free Throws Taken	**Baskets Made**

Jordan	
Free Throws Taken	**Baskets Made**

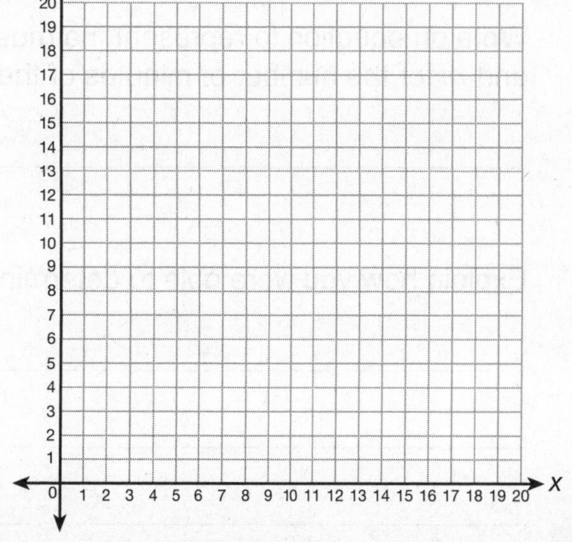

5. A massage therapist gets paid at a rate of $4 for every 5 minutes. Complete the following table of values.

Number of Minutes	Payment (in dollars)

Graph the ordered pairs from the table of values in the following coordinate plane.

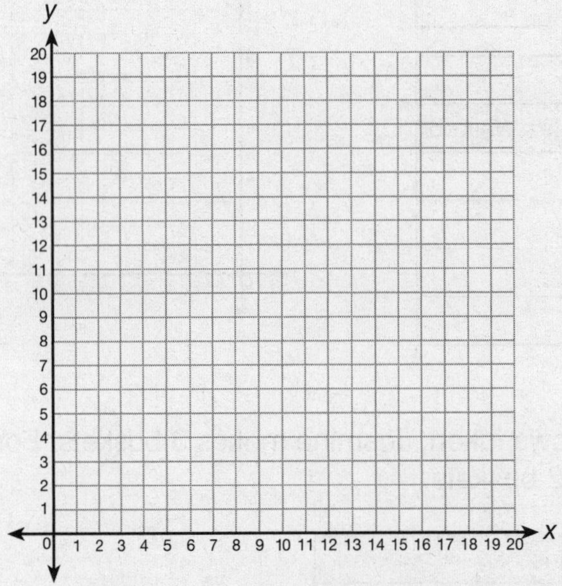

Write an equation to represent the massage therapists' pay. Use t for total payment and m for the number of minutes of therapy.

____

Explain how you were able to determine an equation from the scenario.

____

____

____

CCSS: 6.RP.3.d

Lesson 15: Converting Measurements

You can use ratios to convert from one unit to another in either the metric or U.S. customary system.

There are a few important prefixes that will help you understand the differences between metric units.

kilo-	hecto-	deka-	deci-	centi-	milli-
thousands	hundreds	tens	tenths	hundredths	thousandths

The values of metric units are based on powers of 10. The values of units in the U.S. customary system vary.

Length

The following tables show the units of length in order from smallest to largest. They also show the relationships between units of the same system.

Metric	Conversion
millimeter (mm) about the thickness of a penny	1 mm = $\frac{1}{10}$ cm
centimeter (cm) about the radius of a nickel	1 cm = 10 mm
meter (m) about the height of a kitchen table	1 m = 100 cm 1 m = 1000 mm
kilometer (km) about the length of a 15-minute walk	1 km = 1000 m

U.S. Customary	Conversion
inch (in.) about the diameter of a quarter	1 in. = $\frac{1}{12}$ ft
foot (ft) about the length of a spaghetti noodle	1 ft = 12 in.
yard (yd) about the length of a baseball bat	1 yd = 3 ft 1 yd = 36 in.
mile (mi) about the length of a 20-minute walk	1 mi = 1760 yd 1 mi = 5280 ft

Mass and Weight

The following tables show the units of mass and weight in order from smallest to largest. They also show the conversion relationships between units of the same system. Remember, metric units measure mass, while U.S. customary units measure weight.

Metric	Conversion
milligram (mg) about the weight of a wing of a housefly	1 mg = $\frac{1}{1000}$ g
gram (g) about the weight of a paper clip	1 g = 1000 mg
kilogram (kg) about the weight of a dictionary	1 kg = 1000 g

U.S. Customary	Conversion
ounce (oz) about the weight of a slice of bread	1 oz = $\frac{1}{16}$ lb
pound (lb) about the weight of a full can of seltzer	1 lb = 16 oz
ton (T) about the weight of a small car	1 T = 2000 lb

Capacity

The following tables show the units of capacity in order from smallest to largest. They also show the conversion relationships between units of the same system.

Metric	Conversion
milliliter (mL) about what an eyedropper holds	1 mL = $\frac{1}{1000}$ L
liter (L) about what a medium water bottle holds	1 L = 1000 mL
kiloliter (kL) about what a large wading pool holds	1 kL = 1000 L

U.S. Customary	Conversion
teaspoon (tsp)	1 tsp = $\frac{1}{3}$ tbsp
tablespoon (tbsp)	1 tbsp = 3 tsp
fluid ounce (fl oz)	1 fl oz = 2 tbsp 1 fl oz = 6 tsp
cup (c)	1 c = 8 fl oz
pint (pt)	1 pt = 2 c
quart (qt)	1 qt = 4 c 1 qt = 2 pt
gallon (gal)	1 gal = 4 qt 1 gal = 8 pt 1 gal = 16 c

Ratios can be used to convert units of length, weight, or capacity. You must multiply the given number by the ratio that compares the two units.

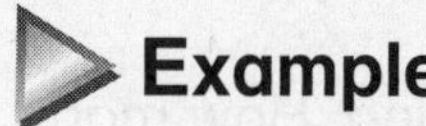

Example

How many feet are in 72 inches?

To solve this problem, you first need to remember the conversion rate from inches to feet. There are 12 inches in 1 foot. One foot, therefore, can be considered as a ratio in terms of inches: $\frac{1}{12}$.

Now you can multiply the total number of inches by this ratio. You will need to convert 72 to $\frac{72}{1}$.

$$\frac{72}{1} \times \frac{1}{12} = \frac{72 \times 1}{1 \times 12} = \frac{72}{12} = 6$$

There are 6 feet in 72 inches.

Example

How many grams are in 5 kilograms?

You need to recall the conversion rate from grams to kilograms. There are 1000 grams in 1 kilogram. A thousand grams, therefore, can be considered as a ratio in terms of one kilogram: $\frac{1000}{1}$.

Now you can multiply the total number of grams by this ratio. You will need to convert 5 to $\frac{5}{1}$.

$$\frac{5}{1} \times \frac{1000}{1} = \frac{5 \times 1000}{1 \times 1} = \frac{5000}{1} = 5000$$

There are 5000 grams in 5 kilograms.

Example

How many cups are in 9 quarts?

There are 4 cups in 1 quart. Four cups, therefore, can be considered as a ratio in terms of one quart: $\frac{4}{1}$.

Now you can multiply the total number of quarts by this ratio. You need to convert 9 to $\frac{9}{1}$.

$$\frac{9}{1} \times \frac{4}{1} = \frac{9 \times 4}{1 \times 1} = \frac{36}{1} = 36$$

There are 36 cups in 9 quarts.

Ratios can be used to convert units of length, weight, or capacity in real-world situations.

Example

Jill and Dina make 3 gallons of lemonade for their lemonade stand. How many cups of lemonade can they sell?

To solve this problem, you first need to remember the conversion rate from gallons to cups. There are 16 cups in 1 gallon. One gallon, therefore, can be considered as a ratio in terms of cups: $\frac{16}{1}$.

Now you can multiply the total number of gallons by this ratio. You will need to convert 3 to $\frac{3}{1}$.

$$\frac{3}{1} \times \frac{16}{1} = \frac{3 \times 16}{1 \times 1} = \frac{48}{1} = 48$$

There are 48 cups in 3 gallons. Therefore, Jill and Dina can sell 48 cups of lemonade.

Example

The height of the roof of the CN Tower in Toronto is 500 yards. The height of the roof of the Petronas Towers in Malaysia is 1242 feet. Joey wants to compare the heights of the structures by converting the height of the Petronas Towers to yards. How many yards tall are the Petronas Towers?

There are 3 feet in 1 yard. One foot, therefore, can be considered as a ratio in terms of yards: $\frac{1}{3}$.

Now you can multiply the total number of feet by this ratio. You will need to convert 1242 to $\frac{1242}{1}$.

$$\frac{1242}{1} \times \frac{1}{3} = \frac{1242 \times 1}{1 \times 3} = \frac{1242}{3} = 414$$

There are 414 yards in 1242 feet. Therefore, the height of the roof of the Petronas Towers is 414 yards. It is lower than the roof of the CN Tower.

TIP: Check that your conversion correctly makes the units smaller or larger. If not, you may have reversed the ratio.

CCSS: 6.RP.3.d

Practice

Directions: For questions 1 through 14, convert the measurements.

1. 4 ft = ____________ in.

2. 80 m = ____________ cm

3. 80 oz = ____________ lb

4. 6000 g = ____________ kg

5. 3 L = ____________ mL

6. 15 tsp = ____________ tbsp

7. 5 gal = ?
   A. 10 pt
   B. 20 pt
   C. 40 pt
   D. 80 pt

8. 180 yd = ____________ ft

9. 9 km = ____________ m

10. 2 T = ____________ lb

11. 4 g = ____________ mg

12. 7 kL = ____________ L

13. 32 fl oz = ____________ c

14. 2 mi = ?
   A. 3520 ft
   B. 5280 ft
   C. 10,460 ft
   D. 10,560 ft

15. Mariano buys a 4-pound bag of trail mix. How many ounces of trail mix did he buy?

    ____

16. A blue whale was once recorded with a length of 36 yards. How many feet long was the whale?

    ____

17. A medical supplier creates 40 liters of a vaccine. How many milliliters did the supplier create?

    ____

18. The maximum weight for a truck in Oregon without a special permit is 40 tons. What is the greatest number of pounds that a truck can weigh in Oregon without a special permit?

    ____

19. The mass of a monitor that was sold at a computer store is listed as 5000 grams. How many kilograms is the mass of the monitor?

    ____

20. Clara's height is 140 centimeters. How many millimeters tall is Clara?

    ____

21. A bottle of vanilla extract contains 4 fluid ounces. How many teaspoons of vanilla extract does the bottle contain?

    ____

    Explain how you found your answer.

    ____

    ____

    ____

Unit 2 Practice Test

Read each question. Choose the correct answer.

1. How many inches are in 3 feet?

   A. 1

   B. 36

   C. 98

   D. 108

2. Ms. Franklin pays $420 for 140 gallons of heating oil. Which description represents her unit rate for the cost of heating oil?

   A. For each gallon Ms. Franklin buys, she pays $420.

   B. For each gallon Ms. Franklin buys, she pays $3.

   C. For each 3 gallons Ms. Franklin buys, she pays $3.

   D. For each 3 gallons Ms. Franklin buys, she pays $1.

3. What is 20% of 50?

   A. 5

   B. 10

   C. 20

   D. 100

4. The following drawing shows the different picks in a guitarist's pocket.

   What is the ratio of black picks to the total number of picks in the guitarist's pocket?

   A. 2:3

   B. 4:2

   C. 3:5

   D. 2:4

5. Jerome burned 75 calories in 15 minutes on a stationary bicycle. How many calories will he burn in 25 minutes?

   A. 5

   B. 85

   C. 125

   D. 140

6. How many millimeters are in 400 centimeters?

   A. 4

   B. 40

   C. 410

   D. 4000

7. Anita read 24 pages of her book in 12 minutes. How long will it take her to read 34 pages?

____

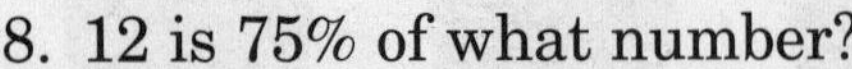

8. 12 is 75% of what number?

____

9. The following drawing shows all the coins under a couch cushion.

What is the ratio of nickels to dimes under the couch cushion?

____

10. A hardware store charges $18 for a 6-pack of energy-efficient light bulbs. What is the unit rate for the cost of 1 energy-efficient light bulb?

____

11. Of all the spectators at Jake's baseball game, 60% are rooting for his team. If there are 120 spectators at Jake's baseball game, how many are rooting for Jake's team?

____

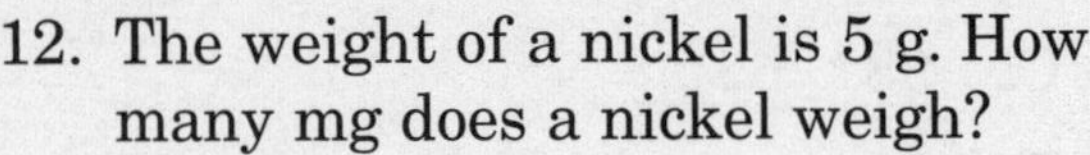

12. The weight of a nickel is 5 g. How many mg does a nickel weigh?

____

13. The following drawing shows the different balls in a duffel bag.

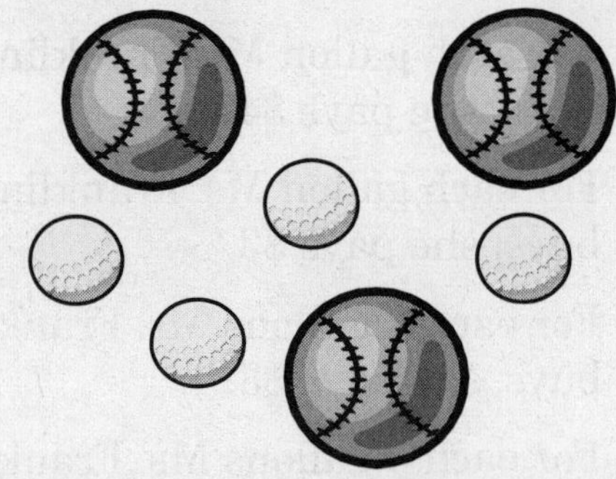

Describe the relationship of golf balls to balls in the sports bag using ratio language.

____

____

14. A pool with 8,000 L of water holds how many kL of water?

____

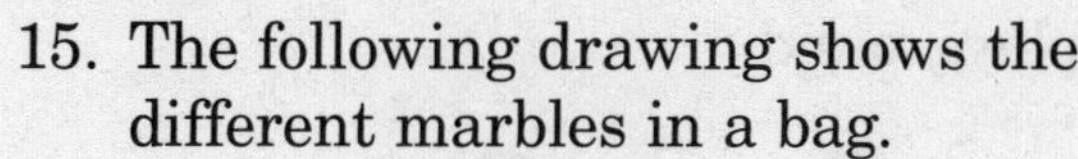

15. The following drawing shows the different marbles in a bag.

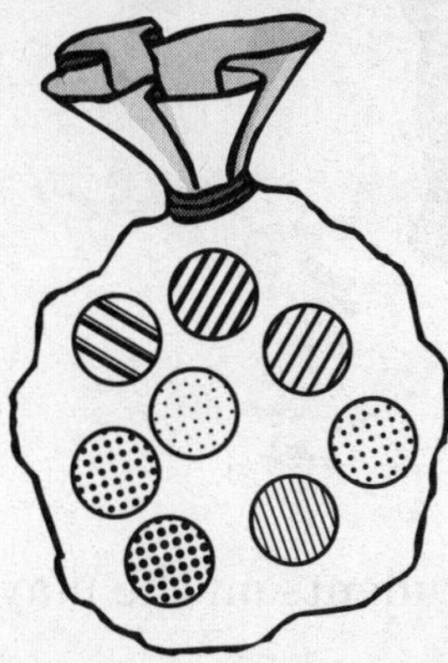

What is the ratio of striped marbles to polka-dotted marbles in the bag?

____

16. A truck has a weight of 4000 pounds. How many tons does the truck weight?

____

17. A brand of diet chips has 30% of the calories of the regular brand. If the diet brand has 90 calories, how many calories does the regular brand have?

____

18. Manuel travels 27 miles in 3 hours on his inline skates. What is his unit rate in miles per hour?

____

19. Ruth makes $35 working 5 hours at her job. How much will she make during an 8-hour workday?

____

20. Several students played together on a school playground.

What is the ratio of students with glasses to all the students on the playground?

____

21. For every 2 minutes in the pool, Jacob can swim 2 laps. For each 2 minutes in the pool, Bijou can swim 3 laps. Complete the following table to create a set of coordinate pairs.

Jacob	
Minutes, *x*	**Laps Swum, *y***

Bijou	
Minutes, *x*	**Laps Swum, *y***

Graph the ordered pairs for each relationship in the following coordinate plane.

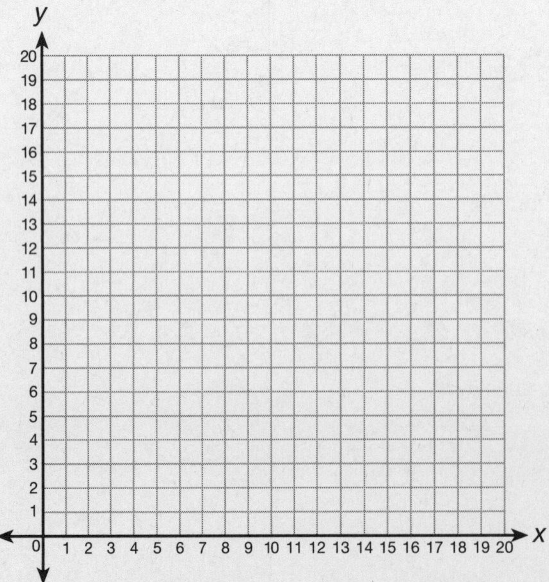

22. A yo-yo rotates 12 times every 3 seconds. What is the yo-yo's unit rate for rotations per second?

____

23. What is 90% of 20?

____

24. How many feet are in 30 yards?

____

25. After 45 seconds, Camden's bathtub fills with 3 inches of water. How many seconds will it take to fill Camden's tub with 20 inches of water?

____

26. A hiker buys a 3-gallon jug of water before going camping. She uses the water to fill several 1-quart water bottles. How many water bottles can she fill?

____

27. Dave draws the following shapes in his notebook.

Describe the ratio of stars to all the shapes in Dave's notebook.

____

28. A principal measures the size of the classrooms in her school. She then uses the equation $y = \frac{1}{5}x$ to determine the greatest number of students who should be allowed in each classroom, where x is the number of square meters and y is the number of students for the room.

    The principal uses the following table of values to determine the number of students for different classrooms. Graph the principal's values in the following coordinate plane.

Number of Square Meters, *x*	Maximum Number of Students, *y*
25	5
50	10
75	15
100	20

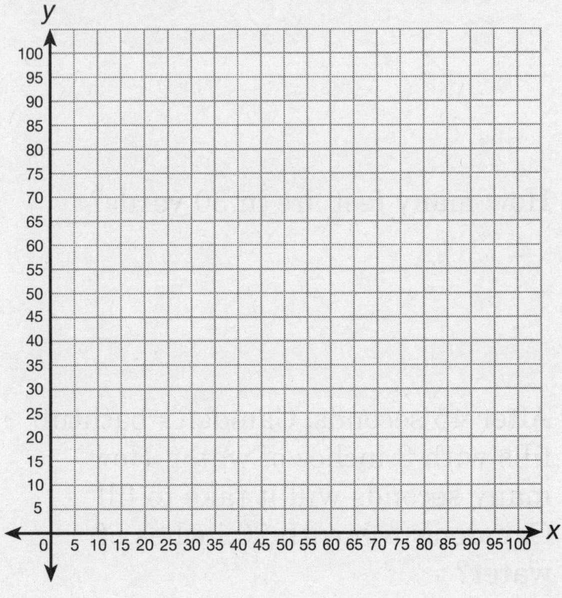

29. During Susie's barbeque, her grill has hot dogs and veggie burgers, as shown.

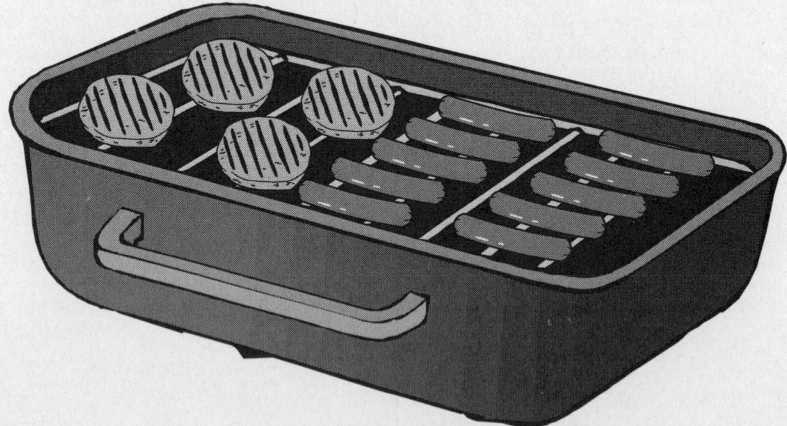

What is the ratio of veggie burgers to hot dogs on the grill?

____

30. While filling his truck with gas, Mr. Dempsey takes the following picture of the gas pump.

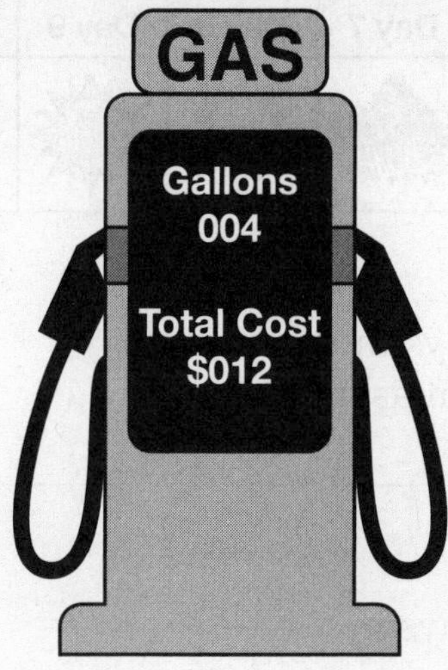

Part A
What is the unit rate of the cost of the gas per gallon?

____

Part B
Mr. Dempsey's gas tank holds 18 gallons of gas. What will be the total cost to fill the empty tank?

____

Part C
Mr. Dempsey fills a 2-gallon container to use with his gas-powered leaf blower. His leaf blower can hold 1 quart of gas. How many times can Mr. Dempsey fill his leaf blower from the 2-gallon container?

____

Explain how you found the number of times Mr. Dempsey could fill his leaf blower.

____

____

____

____

31. A weather station's 10-day forecast is shown below.

Weather Station's 10-Day Forecast

Day 1	Day 2	Day 3	Day 4	Day 5	Day 6	Day 7	Day 8	Day 9	Day 10

Part A
What is the ratio of sunny days to cloudy days in the forecast? Write your answer in simplest form as a fraction.

____

Part B
What is the ratio of cloudy days to all the days in the forecast? Write your answer in simplest form using a colon (:).

____

Part C
Describe the relationship of cloudy days to sunny days using a ratio. Simplify the numbers if possible.

____

____

Explain how you knew whether you could simplify the numbers or not.

____

____

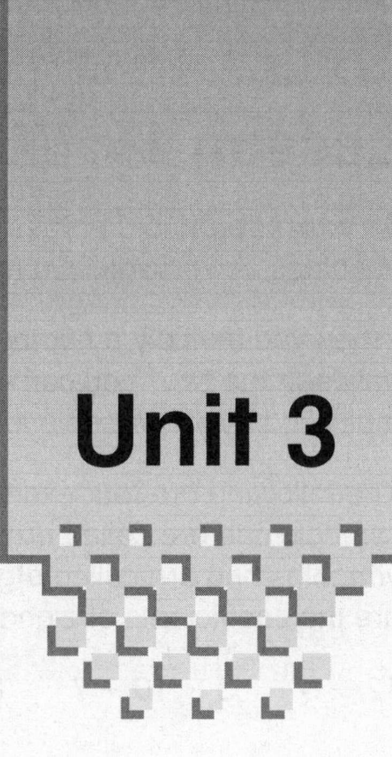

Expressions and Equations

It may seem strange at first to see letters, or variables, in math problems. But the truth is that you use variables all the time in real life. Sometimes, you may not even realize it. For example, pretend that you're reading a book that is 80 pages long, and you've read 55 pages. To figure out how many pages are left, you can add the 55 pages to some number to get 80. You can write "some number" as a letter. Those letters, or variables, are useful for many different situations in math. This unit will show you some of them.

In this unit, you will learn many of the ways that variables can appear in a math problem. You will learn to write expressions, inequalities, and equations with variables. You will also learn to find solutions for those expressions, inequalities, and equations. You will identify and create equivalent expressions using properties of numbers. Finally, you will learn the difference between dependent and independent variables and create tables to show their relationships.

In This Unit

Lesson 16: Writing Expressions

An **expression** is a phrase made up of numbers, operation symbols, and sometimes variables. A **variable** is a letter that represents an unknown number.

When you multiply a number by a variable, you do not need a multiplication sign between the two. You can write the number directly in front of the variable. This number multiplied by the variable is called the **coefficient.**

The following are four examples of expressions. The first two contain only numbers and symbols and are called **numerical expressions.** The last two expressions have variables and are called **algebraic expressions.** In the algebraic expressions, 8 and 4 are the coefficients of *z* and *y*.

$12 \div 3$ $3 \times 10 + 7$ $8z - 5$ $z + 4y$

variables

A **term** is a part of an expression that is either a number, a variable, or the product of a number and a variable. In the expression $2x^2 + 8y - 5$, there are three terms: $2x^2$, $8y$, and 5. The operations are not part of the terms. **Like terms** have the same variable.

To translate a word expression into a mathematical expression, you need to identify the operations. Here are some related key words or phrases for addition, subtraction, multiplication, division, and exponents.

Addition: sum, more, more than, plus, increased by, gain

Subtraction: difference, less, less than, minus, decreased by, loss

Multiplication: product, multiplied by, times, double, triple

Division: quotient, divided by, per, ratio, half, third, fourth, into, equal groups

Exponents: times itself, squared, cubed

TIP: Keep in mind that some words are used for more than one operation in an expression. For example, the word *factor* can represent division (the factors of 10 are 1, 2, 5, and 10). It can also show multiplication by representing a number that is multiplied by another number. (The factors in 3×4 are 3 and 4.) You need to consider the context of the word problem.

CCSS: 6.EE.1, 6.EE.2.a, 6.EE.2.b, 6.EE.6

You can translate words into a numerical or algebraic expression. When the expression includes either addition or multiplication only, the order in which the terms are written **does not matter.** (This is because the operations of addition and multiplication are **commutative.**)

Example

Write a numerical expression to represent "four plus three times itself."

The word *plus* indicates **addition.** The words *times itself* indicate an exponent.

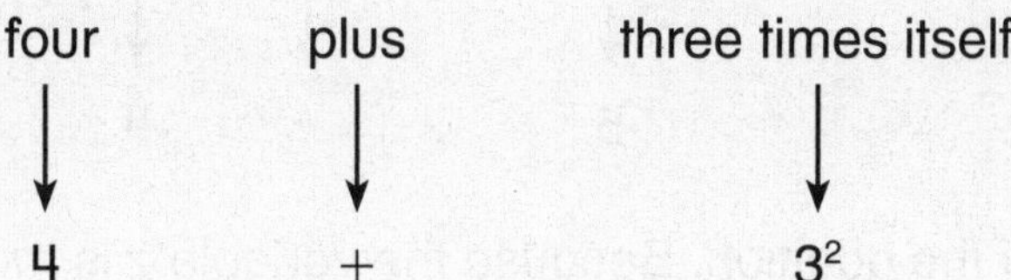

The expression can be written as $4 + 3^2$. It can also be written as $3^2 + 4$.

When an expression includes subtraction or division, the order in which the terms are written **does matter.**

Example

Write an algebraic expression to represent "six less than twice a number." Let $z =$ the number.

The phrase *less than* indicates **subtraction.** The two terms in the expression are 6 and $2z$.

Which of the following expressions is correct: $6 - 2z$ or $2z - 6$?

The phrase *less than* indicates that you have to change the order of the terms in the expression from the way they appear in the description.

The expression "six less than twice a number" can only be written as $2z - 6$.

Expressions can be represented using the distributive property.

Example

Write an algebraic expression to represent "a number multiplied by the sum of five and four." Let n = the number.

The words *multiplied by* indicate **multiplication.** The word *sum* indicates **addition.**

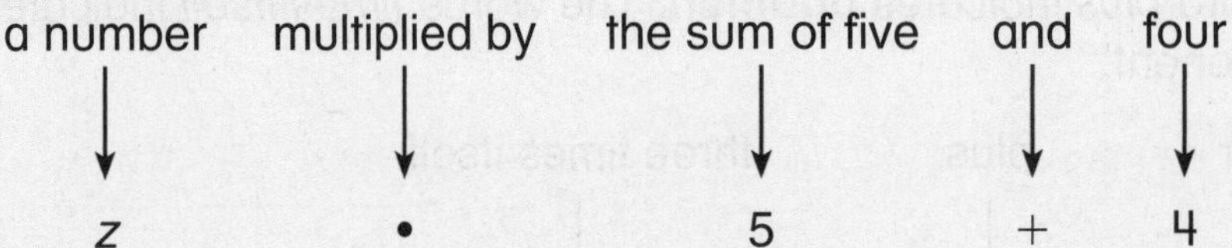

The sum is the result of the addition. Because the variable z is multiplied by the *sum of* 5 and 4, you must add before you can multiply. Parentheses can be used to separate the addition from the rest of the expression. The expression can be written as $z \cdot (5 + 4)$.

You do not need a multiplication sign between the variable and the expression in parentheses. Therefore, the expression can be written as $z(5 + 4)$.

Example

Write an expression to represent "six times the sum of seven plus three."

The word *times* indicates **multiplication.** The words *sum of* and *plus* indicate **addition.**

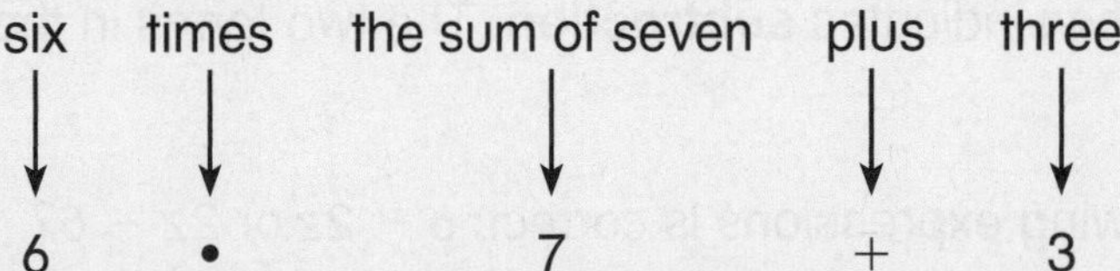

Because 6 is multiplied by the *sum of* 7 and 3, you must first add before you can multiply. Use parentheses to separate the addition. The expression can be written as $6 \cdot (7 + 3)$, or $6(7 + 3)$.

The factor 6 can be multiplied by the combined sum of 7 and 3: $6(7 + 3)$, or $6 \cdot 10$. By the distributive property, it can also be multiplied by the 7 and 3 factors independently: $6 \cdot 7 + 6 \cdot 3$.

Practice

Directions: For questions 1 through 12, write a numerical or algebraic expression from the words. Let n = the number.

1. eleven more than two

   ____

2. nine less than twelve

   ____

3. the product of two and three

   ____

4. sixteen divided by two times itself three times

   ____

5. forty divided by the sum of one and nine

   ____

6. three times the sum of four and seven

   ____

7. a number plus ten

   ____

8. eight less than a number squared

   ____

9. twenty divided by a number

   ____

10. the cube of a number and five

    ____

11. three less than the product of a number and two

    ____

12. eleven more than six divided by a number

    ____

Directions: For questions 13 through 17, identify the underlined part of the expression. Write *sum, term, product, factor, quotient, variable,* or *coefficient.*

13. $4x + \underline{3y} + 2$

____

14. $\underline{17}z - 8$

____

15. $\underline{4}(b + 1)$

____

16. $8 \bullet 7\underline{n}$

____

17. $\underline{3x^2} \div 99$

____

Directions: For questions 18 through 21, write an algebraic expression to represent each situation.

18. Carissa divided 40 grapes equally among *f* friends. How many grapes did each friend get?

____

19. Jermaine has 5 times as many albums as Miguel, *m*. How many albums does Jermaine have?

____

20. This week, Olive got 5 dollars less than her usual allowance, *a*. How much did Olive get for her allowance this week?

____

21. Patrick ran 6 kilometers. Then he ran *k* more kilometers. How many kilometers did he run in total?

____

22. Nadine counted 25 people on her train when she got on. At the next stop, *m* more people got on the train, and 6 people got off. Write an expression that describes how many people were on Nadine's train after the next stop.

____

Explain how you found your answer.

____

____

CCSS: 6.EE.1, 6.EE.2.c

Lesson 17: Evaluating Expressions

To evaluate an expression with a variable or symbols, substitute the given number for the variable or symbol. Then follow the order of operations to simplify the expression. If there is more than one variable or symbol, substitute each of their values with the given numbers.

Example

Evaluate $k + 25$ for $k = 16$.

Given that $k = 16$, substitute 16 for k in the expression and simplify.

$k + 25 = (16) + 25$	**Substitute 16 for *k*.**
$= 41$	**Add.**

The expression $k + 25$ evaluated for $k = 16$ is 41.

Example

Evaluate $8y - 3$ for $y = 0.9$.

Given that $y = 0.9$, substitute 0.9 for y in the expression, and simplify.

$8y - 3 = 8(0.9) - 3$	**Substitute 0.9 for *y*.**
$= 7.2 - 3$	**Multiply 8 and 0.9.**
$= 4.2$	**Subtract.**

The expression $8y - 3$ evaluated for $y = 0.9$ is 4.2.

Example

Evaluate $4m + 3n^2$ for $m = 3$ and $n = 5$.

Given that $m = 3$ and $n = 5$, substitute 3 for m and 5 for n in the expression, and simplify.

$4m + 3n = 4(3) + 3(5)^2$	**Substitute 3 for *m* and 5 for *n*.**
$= 12 + 75$	**Multiply 4 by 3 and 3 by 5.**
$= 87$	**Add.**

The expression $4m + 3n$ evaluated for $m = 3$ and $n = 5$ is 87.

You can evaluate an expression to solve a formula. Substitute the given values into the expression on one side of the equal sign.

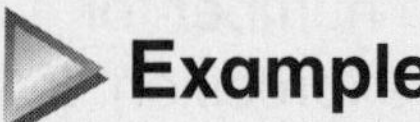

Example

The formula to convert temperature from Celsius to Fahrenheit is $F = \frac{9}{5}C + 32$. What is the Fahrenheit temperature, *F*, if the Celsius temperature, *C*, is 25^0?

Given that $C = 25$, substitute 25 for *C*, and simplify.

$F = \frac{9}{5}C + 32$	
$F = \frac{9}{5}(25) + 32$	**Substitute 25 for *C*.**
$F = 45 + 32$	**Multiply $\frac{9}{5}$ and 25.**
$F = 77$	**Add.**

If the Celsius temperature, *C*, is 25^0, the Fahrenheit temperature, *F*, is 77^0.

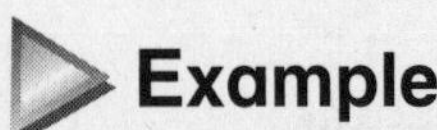

Example

The formula for the area of a triangle is $A = \frac{1}{2}bh$, where *A* is the area, *b* is the base of the triangle, and *h* is the height of the triangle. What is the area of a triangle with a base of 8.8 and a height of 10.5?

Given that $b = 8.8$ and $h = 10.5$, substitute 8.8 for *b* and 10.5 for *h*, and simplify.

$A = \frac{1}{2}bh$	
$A = \frac{1}{2}(8.8)(10.5)$	**Substitute 8.8 for *b* and 10.5 for *h*.**
$F = \frac{1}{2}(92.4)$	**Multiply 8.8 and 10.5.**
$F = 46.2$	**Multiply $\frac{1}{2}$ and 92.4.**

The area of the triangle is 46.2 square units.

Practice

Directions: For questions 1 through 7, evaluate each expression for $y = 4$ and $z = 7$.

1. $15 + z$ ________________________

2. $4 + (z + 10)$ ________________________

3. $25 - 2y$ ________________________

4. $30 - z$ ________________________

5. $8y^2$ ________________________

6. $\frac{z + 18}{5}$ ________________________

7. $2y - z$

   A. 1

   B. 7

   C. 8

   D. 10

Directions: For questions 8 through 14, evaluate each expression for $a = 5$, $b = 2$, and $c = 4$.

8. $4c \div 8 + 2a$ ________________________________

9. $25 + a + 6 - a$ ________________________________

10. $4a - 19 + 5c$ ________________________________

11. $(7 - c) \bullet b^3$ ________________________________

12. $13c + 23 - 3a$ ________________________________

13. $a(c \div b)$ ________________________________

14. $2a + 20$

    A. 10

    B. 15

    C. 25

    D. 30

Directions: For questions 15 through 19, evaluate the expressions in the formulas with the given values.

15. The volume of a cube is $V = s^3$, where V is the volume and s is the length of one side. What is the volume of a cube if the length of a side, s, is 10 ft?

 ____

16. The area of a parallelogram is $A = bh$, where b is the base and h is the height of the parallelogram. What is the area of a parallelogram with a base of 5 cm and a height of 20 cm?

 ____

17. The volume of a rectangular prism is $V = lwh$, where l is the length, w is the width, and h is the height of the prism. What is the volume of a rectangular prism with a length of 5 in. and a width and height of 6 in. each?

 ____

18. A taxi driver determines the fare using the formula $F = 1.50 + 2.6m$, where F is the total fare and m is the number of miles driven. What is the fare for a taxi ride that is 2.5 miles long?

 ____

19. A waiter determines his salary for a week using the formula $S = 3.75h + t$, where S is his salary, h is the number of hours he works, and t is his total tips. How much does the waiter make if he works 30 hours in a week and makes a total of $538.75 in tips?

 ____

 Explain how you found your answer.

 ____

 ____

 ____

Lesson 18: Equivalent Expressions

Two expressions may look different even though they represent the same information or value. These expressions are said to be **equivalent.** You can use the commutative, associative, and distributive properties to find out whether two expressions are equivalent.

Example

Are the following two expressions equivalent?

$5(x + 3) \qquad 5x + 15$

Using the distributive property, rewrite one of the expressions.

$$5(x + 3) = 5 \bullet x + 5 \bullet 3$$
$$= 5x + 15$$

Because $5(x + 3) = 5x + 15$, the expressions are equivalent.

Two expressions are also equivalent if you substitute the same value for the same variables and both expressions are equal.

Example

Determine whether $(x + 3) + 4$ is equivalent to $x + 8$ by letting $x = 3$ and $x = -2$.

When $x = 3$:

$$(x + 3) + 4 = ((3) + 3) + 4 \qquad x + 8 = (3) + 8$$
$$= (6) + 4 \qquad = 11$$
$$= 10$$

When $x = -2$:

$$(x + 3) + 4 = ((-2) + 3) + 4 \qquad x + 8 = (-2) + 8$$
$$= (1) + 4 \qquad = 6$$
$$= 5$$

For both of these x-values, the two expressions don't agree. The expressions $(x + 3) + 4$ and $x + 8$ are not equivalent.

CCSS: 6.NS.4, 6.EE.3, 6.EE.4

You can often simplify an expression to create an equivalent expression. You may need to use the order of operations.

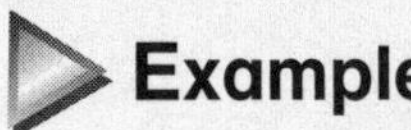

Example

Simplify the following expression to create an equivalent expression.

$3(x + 12) + 2x - 11$

To simplify the expression, you can use the distributive property. Then you can combine the like terms. Be sure to follow the order of operations.

$3(x + 12) + 2x - 11$	
$3 \bullet x + 3 \bullet 12 + 2x - 11$	**Use the distributive property.**
$3x + 36 + 2x - 11$	**Perform the multiplication first.**
$5x + 25$	**Combine the like terms.**

Because $3(x + 12) + 2x - 11 = 5x + 25$, the expressions are equivalent.

Example

Simplify the following expression to create an equivalent expression.

$4(\frac{3}{2}x + 2) \div 2 - 2^2 \bullet x$

To simplify the expression, you can use the distributive property. Then you can combine the like terms. Be sure to follow the order of operations.

$4(\frac{3}{2}x + 2) \div 2 - 2^2 \bullet x$	
$4(\frac{3}{2}x + 2) \div 2 - 4 \bullet x$	**Simplify the exponents first.**
$(4 \bullet \frac{3}{2}x + 4 \bullet 2) \div 2 - 4 \bullet x$	**Use the distributive property.**
$(6x + 4 \bullet 2) \div 2 - 4 \bullet x$	**Multiply/divide from left to right.**
$(6x + 8) \div 2 - 4 \bullet x$	**Multiply/divide from left to right.**
$3x + 4 - 4 \bullet x$	**Multiply/divide from left to right.**
$3x + 4 - 4x$	**Multiply/divide from left to right.**
$-x + 4$	**Combine like terms.**

$4(\frac{3}{2}x + 2) \div 2 - 2^2 \bullet x$ is equivalent to $-x + 4$.

TIP: The order of operations states that all operations within **P**arentheses are solved first, followed by **E**xponents. **M**ultiplication and **D**ivision are then solved from left to right, followed by **A**ddition and **S**ubtraction from left to right. These operations can be abbreviated as PEMDAS.

Practice

Directions: For questions 1 through 6, determine whether the given expressions are equivalent. Write *yes* if they are and *no* if they are not.

1. $2a + 3 \cdot (1 + 2)$ $\quad$ $2a + 9$ ________________

2. $5(3 + 2) + x$ $\quad$ $2(6 + x) + 10$ ________________

3. $4 + (z + 10)$ $\quad$ $4 + (10 + z)$ ________________

4. $3a + (19 + 2z)$ $\quad$ $(3a + 19) + 2z$ ________________

5. $33y$ $\quad$ $(3 \cdot y) \cdot 11$ ________________

6. $4(3b - 4)$ $\quad$ $2(6b - 8)$ ________________

7. Which expression is equivalent to the following expression?

   $$6x + 4(1 + x) + 5$$

   A. $6x + 9$
   B. $6x + 10$
   C. $10x + 9$
   D. $10x + 10$

8. Which expression is equivalent to the following expression?

   $$4y \cdot 2(2 + 1)$$

   A. $4y + 5$
   B. $4y + 6$
   C. $20y$
   D. $24y$

Directions: For questions 9 through 15, simplify the given expressions.

9. $10 + 2(b - 4) - b$ ________________

10. $3 + z + z + z + z + z + 2$ ________________

11. $10 + 4x + x(3 + 2) - 1$ ________________

12. $3x + 2x \div 4 - 2$ ________________

13. $3k - 4 + k + 8 - 2k - 1$ ________________

14. $10(2b + 1) - 4(3b + 2)$ ________________

15. $5(7m + 3 - 2m - 2) - 4m + n - 4$ ________________

    Explain how you found your answer.

    ____

    ____

    ____

Lesson 19: Solving Equations

An **equation** is a mathematical statement that has an equal sign, =. The equal sign separates two expressions and shows that the expressions are equal. Remember, an expression does not contain an equal sign. Different examples of equations are shown below.

$12 + 13 = 25$ $\quad$ $10 + x = 15$ $\quad$ $2x + 4 = 3x + 1$

Many equations will have a variable on one or both sides. Solving an equation means finding the value for the variable that makes the expressions equal to each other. To solve an equation, you need to isolate the variable (get it alone on one side of the equation). Then you can use inverse operations to find the value of the variable.

You can check a solution by substituting it for the variable in the original equation.

Example

A solution to the following equation is $n = 3$. Is the solution correct?

$3n + 4 = 13$

To check the solution, substitute the value for n into the original equation.

$3n + 4 = 13$	
$3(3) + 4 \stackrel{?}{=} 13$	**Substitute 3 for *n*.**
$9 + 4 \stackrel{?}{=} 13$	**Multiply 3 by 3.**
$13 \stackrel{?}{=} 13$	**Add.**

Because $13 = 13$, the solution $n = 3$ is correct.

Example

A solution to the following equation is $p = 5$. Is the solution correct?

$8p = 40$

To check the solution, substitute the value for p into the original equation.

$8p = 40$	
$8(5) \stackrel{?}{=} 40$	**Substitute 5 for *p*.**
$40 \stackrel{?}{=} 40$	**Multiply 8 by 5.**

Because $40 = 40$, the solution $p = 5$ is correct.

TIP: Not every equation has a solution. There may be no solution to an equation, such as $x + 1 = x$.

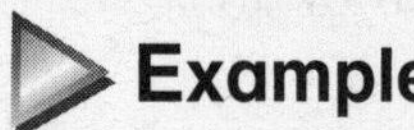

CCSS: 6.EE.5, 6.EE.6, 6.EE.7

You can solve equations by using inverse operations to create equivalent equations. If two equations are equivalent, they have the same solution.

Example

Solve the following equation for x. Then check the answer.

$4 + x = 15$

Use inverse operations to get the variable alone on one side of the equation.

$4 + x = 15$
$4 + x - 4 = 15 - 4$ **Subtract 4 from both sides.**
$x = 11$

The solution to the equation is $x = 11$. To check the solution, substitute the value for the variable in the original equation.

$4 + x = 15$
$4 + (11) \stackrel{?}{=} 15$ **Substitute 11 for *x*.**
$15 \stackrel{?}{=} 15$ **Add.**

Because $15 = 15$, the solution is correct. The value of x is 11.

The solution to $x = 11$ is 11. Because $4 + x = 15$ and $x = 11$ are equivalent equations, the solution to $4 + x = 15$ is also 11.

Example

Solve the following equation for y. Then check the answer.

$8y = 72$

Use inverse operations to get the variable alone on one side of the equation.

$8y = 72$
$8y \div 8 = 72 \div 8$ **Divide both sides by 8.**
$y = 9$

The solution to the equation is $y = 9$. To check the solution, substitute the value for the variable in the original equation.

$8y = 72$
$8(9) \stackrel{?}{=} 72$ **Substitute 9 for y.**
$72 \stackrel{?}{=} 72$ **Add.**

Because $72 = 72$, the solution is correct. The value of y is 9.

TIP: Addition and subtraction are inverse operations, as are multiplication and division.

You can write an equation from a given scenario. Translate the numbers and operations as you would when writing expressions. However, be sure to include an equal sign to separate the two expressions. The word *is* or *equals* is often an indicator of an equal sign.

Example

Write an equation to represent the following scenario.

A number increased by nine equals ten. Let n = the number.

Because the variable is *n,* replace "a number" with *n.* The phrase *increased by* represents addition. The word *equals* should be represented by an equal sign.

Now you can write the equation.

A number increased by nine equals ten
n + 9 = 10

The equation that represents the scenario is $n + 9 = 10$.

Example

Write an equation to represent the following scenario.

The product of a number and five is fifteen. Let x = the number.

Replace "a number" with *x.* The word *product* represents multiplication. The word *is* should be represented by an equal sign.

Remember that when a variable is multiplied by a number, you do not need the multiplication symbol. The number becomes the coefficient of the variable. Now write the equation.

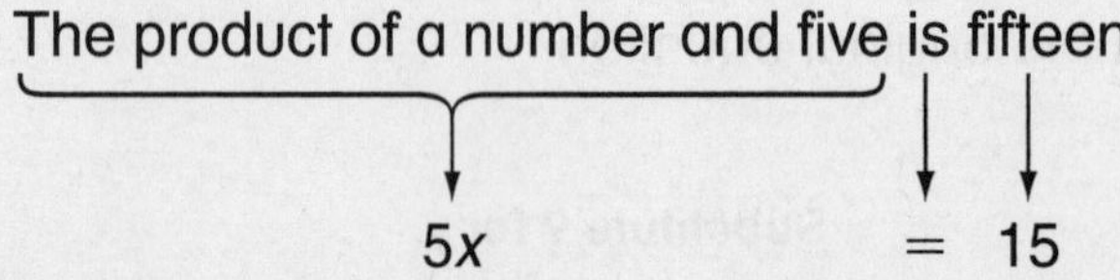
The product of a number and five is fifteen
5x = 15

The equation that represents the scenario is $5x = 15$.

CCSS: 6.EE.5, 6.EE.6, 6.EE.7

You can write an equation to represent a real-world problem situation. Then you can solve your equation to answer the question asked in the problem.

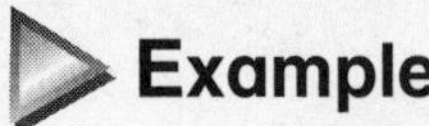

Example

Write an equation to represent the following problem situation. Then solve it.

On a trip from San Diego, CA, to Phoenix, AZ, Billy drove 188 miles. Allison drove the rest of the way. In total, the trip from San Diego to Phoenix was 355 miles. How many miles did Allison drive?

First, list all the information you know.

the number of miles that Billy drove = 188
the number of miles that Allison drove = x
the total number of miles driven = 355

Billy and Allison each drove some number of miles. Combined, they drove a total of 355 miles. Now you can write an equation using addition.

Billy's miles ⟶ $188 + x = 355$ ⟵ **total number of miles**

↑ **Allison's miles**

To solve the equation, isolate the variable x. That means using the inverse operation to get x by itself on one side of the equation. That will tell you its value.

$188 + x = 355$
$188 + x - 188 = 355 - 188$ **Subtract 188 from both sides.**
$x = 167$

Allison drove 167 miles.

TIP: There may be several ways to write an equation to represent a problem situation. For this example, $x + 188 = 355$ would also work, as would $355 - x = 188$ or $355 - 188 = x$.

Practice

Directions: For questions 1 through 6, determine whether the given value is a solution of the equation. Write *yes* if it is and *no* if it is not.

1. $x + 20 = 50$

   $x = 70$ ____________

2. $7y = 63$

   $y = 9$ ____________

3. $5z = 10$

   $z = 5$ ____________

4. $k + 14 = 41$

   $k = 27$ ____________

5. $12m = 60$

   $m = 5$ ____________

6. $22 + n = 23$

   $n = 45$ ____________

Directions: For questions 7 through 12, solve the equation.

7. $33 + z = 51$

   $z =$ ____________

8. $10d = 10$

   $d =$ ____________

9. $b + 90 = 99$

   $b =$ ____________

10. $5h = 55$

    $h =$ ____________

11. $q + 4 = 80$

    $q =$ ____________

12. $18s = 36$

    $s =$ ____________

13. What is the solution for $20w = 100$?

    A. $w = 5$

    B. $w = 50$

    C. $w = 80$

    D. $w = 120$

14. What is the solution to $2 + p = 50$?

    A. $p = 25$

    B. $p = 48$

    C. $p = 52$

    D. $p = 58$

CCSS: 6.EE.5, 6.EE.6, 6.EE.7

Directions: For questions 15 through 20, write an equation to model the real-world scenario. Then solve the equation.

15. A school bus holds 26 students. How many buses are needed to carry 78 students? Let b = the number of buses.

16. Marcus had 34 pictures on his digital camera. After he took some pictures of a sunset, he had 53 pictures on his camera. How many pictures did he take of the sunset? Let p = the pictures of the sunset.

17. Antoine bought several bags of marbles. Each bag has 14 marbles. In total, Antoine bought 56 marbles. How many bags did he buy? Let m = the bags.

18. Monique had a collection of model cars. She got 8 model cars for her birthday. She now has 25 model cars. How many cars did she have in her collection before her birthday? Let c = the number of model cars in her collection.

19. Gail jogged around the track 4 times. She jogged a total of 600 meters. What is the length of the track? Let t = the length of the track.

20. A parking meter had 18 quarters in it by noon. By the end of the day, it had 41 quarters in it. How many quarters were put into the parking meter in the afternoon? Let q = the number of quarters added in the afternoon.

    Explain how you set up the equation and solved for it.

Lesson 20: Solving Inequalities

An **inequality** is a mathematical sentence comparing two expressions that are not equal. An inequality may use one of the following symbols: $<$ or $>$. The symbol $>$ means "is greater than." The symbol $<$ means "is less than."

To solve inequalities, follow the same rules as for solving equations. Use inverse operations to isolate the variable.

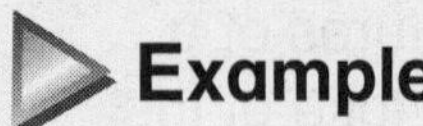

Example

Solve the following inequality for *x*.

$5x < -35$

Use inverse operations to solve for *x*.

$5x < -35$

$\frac{5x}{5} < -\frac{35}{5}$ **Divide both sides by 5.**

$x < -7$

The solution set for the inequality is $x < -7$.

The graph of the solution set is shown below.

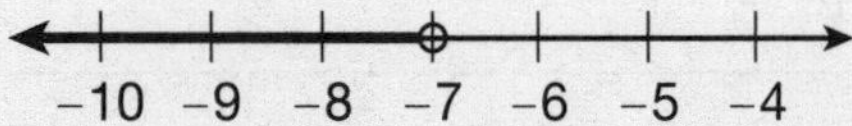

Notice that the dot on -7 is open. This means that -7 is not included as part of the solution set.

To check the answer, substitute any number less than -7 for *x*. Use $x = -8$.

$5x < -35$

$5(-8) < -35$

$-40 < -35$

CCSS: 6.EE.5

Example

Solve the following inequality for *x*.

$8 + x \geq 5$

Use inverse operations to solve for x.

$8 + x \geq 5$
$8 - 8 + x \geq 5 - 8$ **Subtract 8 from both sides.**
$x \geq -3$

The solution set for the inequality is $x \geq -3$.

The graph of the solution set is shown below.

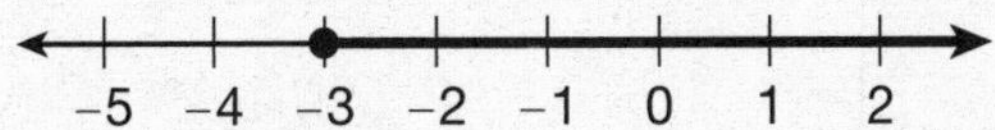

Notice that the dot on −3 is filled in. This means that −3 is included in the solution set.

To check the answer, substitute any number greater than or equal to −3 for *x*. Use $x = 0$.

$8 + x \geq 5$
$8 + (0) \geq 5$
$8 \geq 5$

Practice

Directions: For questions 1 through 10, solve the inequality.

1. $3m > 72$ ____________

2. $\frac{1}{5}x > 30$ ____________

3. $n - 4 \leq 8$ ____________

4. $2n < 5$ ____________

5. $34 + z \geq 14$ ____________

6. $n - 55 > 9$ ____________

7. $n + 90 < -101$ ____________

8. $3y \leq 24$ ____________

9. $x - 49 < 7$ ____________

10. $12x \leq 168$ ____________

CCSS: 6.EE.5

Directions: For questions 11 through 17, solve the inequality for the given variable. Then graph the solution to the inequality on a number line.

11. $3x < 27$

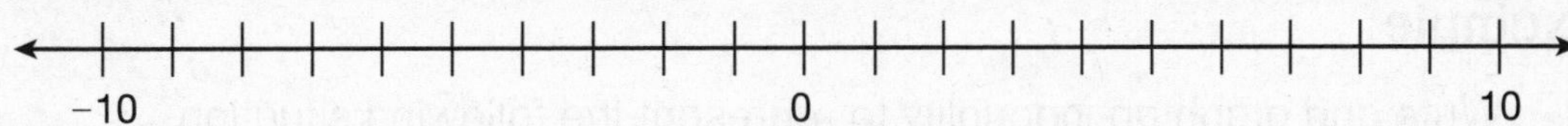

12. $y - 6 \geq 2$

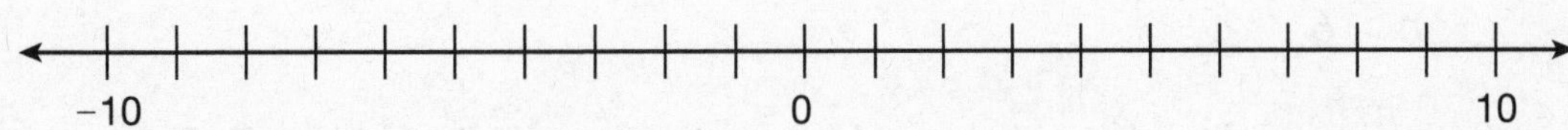

13. $4 + n > 9$

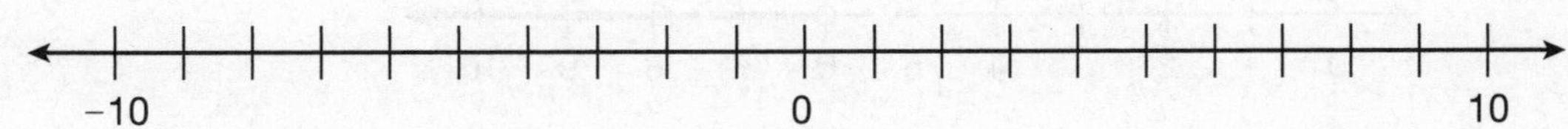

14. $12 \leq x + 5$

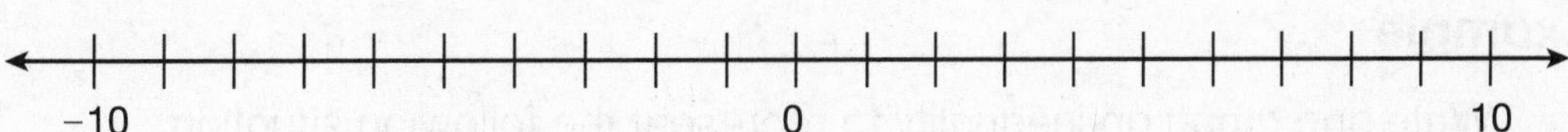

15. $x + 2 \leq -1$

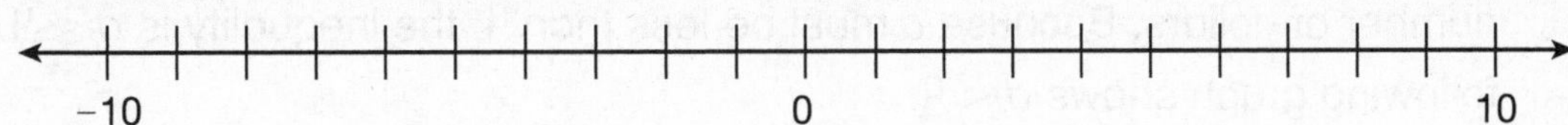

16. $8 > 4n$

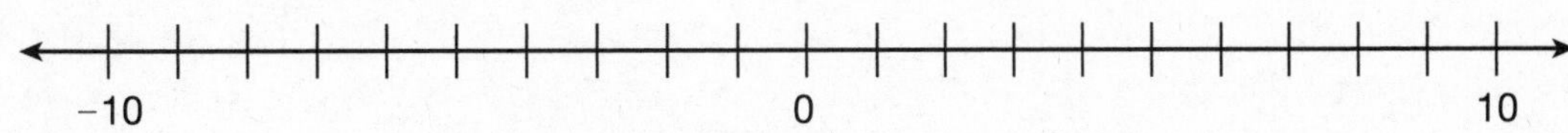

17. $5x \geq -10$

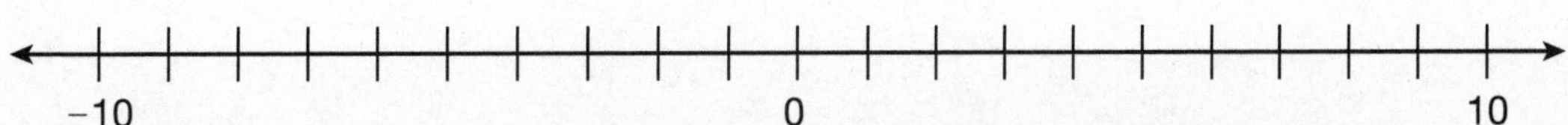

Lesson 21: Writing Inequalities

You can write inequalities that describe a real-world situation.

Example

Write and graph an inequality to represent the following situation.

During the school year, Hermano has to write more than 6 essays for his English class.

$n > 6$

In the inequality above, *n* must be greater than 6. It could be 7, 8, or even 200. There are an infinite number of solutions for the inequality. The following graph shows $n > 6$.

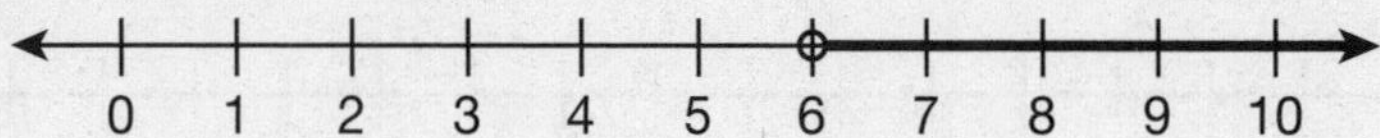

The empty circle shows that 6 is not a solution. The dark arrow to the right shows that Hermano has to write more than 6 essays.

Example

Write and graph an inequality to represent the following situation.

Isaac needs to spend less than $4 on a new notebook.

Use a variable to represent how much Isaac can spend. You can use *d* for the number of dollars. Because *d* must be less than 4, the inequality is $d < 4$. The following graph shows $d < 4$.

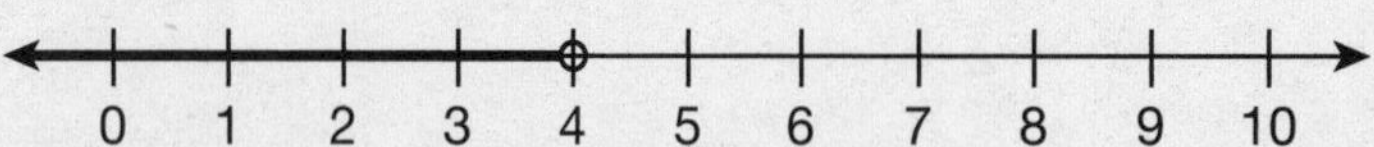

Any number less than 4 would be a solution.

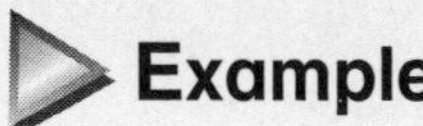

Example

A computer company makes a new laptop that weighs less than 5 pounds. Write an inequality to represent this scenario. Then graph the inequality and write a possible solution.

Use a variable to represent how much the laptop can weigh. You can use p for the number of pounds. The variable p must be less than 5. Therefore, the following inequality represents the scenario.

$p < 5$

To make the graph, draw a circle at 5. Then draw a line that points to the left to show that any value less than 5 is a solution. The following graph shows $p < 5$.

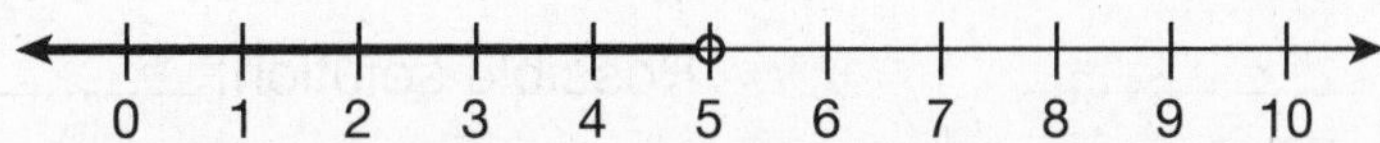

A possible weight for the laptop computer is 4.2 pounds.

Example

A passenger on a roller coaster must be more than 3 feet tall. Write an inequality to represent this scenario. Then graph the inequality and write a possible solution.

Use a variable to represent how tall the person can be. You can use f for the number of feet. The variable f must be greater than 3. Therefore, the following inequality represents the scenario.

$f > 3$

To make the graph, draw a circle at 3. Then draw a line that points to the right to show that any value greater than 3 is a solution. The following graph shows $f > 3$.

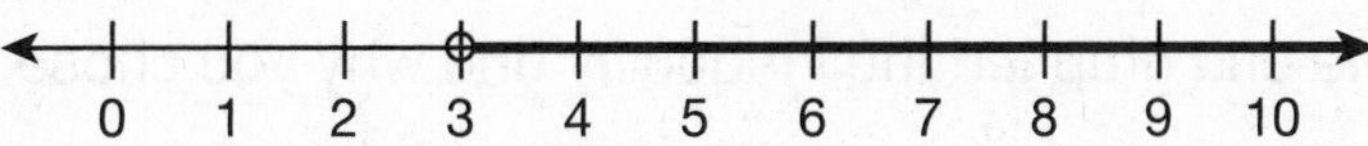

A possible height for a passenger is $3\frac{1}{2}$ feet.

Practice

Directions: For questions 1 through 4, write an inequality to represent the real-life situation. Then complete the graph and write a possible solution.

1. Kids on the baby playground must be less than 5 years old. Let y = the number of years old.

   Inequality: ____________ Possible solution: ____________

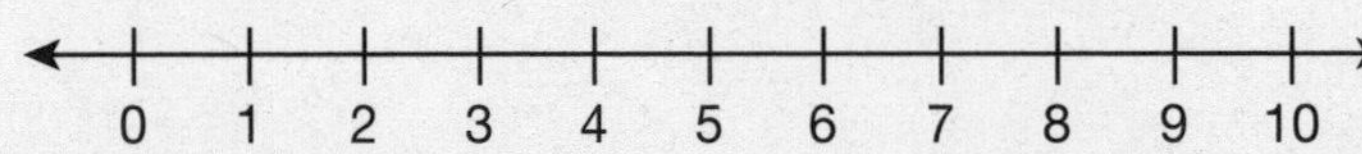

2. Roxanne needs more than 8 tickets to get a stuffed animal at the fair. Let c = the number of tickets needed.

   Inequality: ____________ Possible solution: ____________

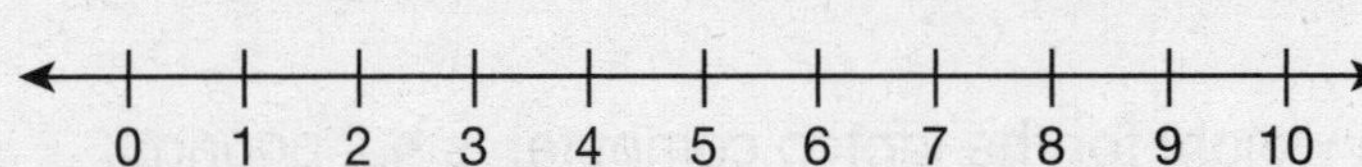

3. More than 2 buses are needed to bring all the students to the aquarium. Let b = the number of buses needed.

   Inequality: ____________ Possible solution: ____________

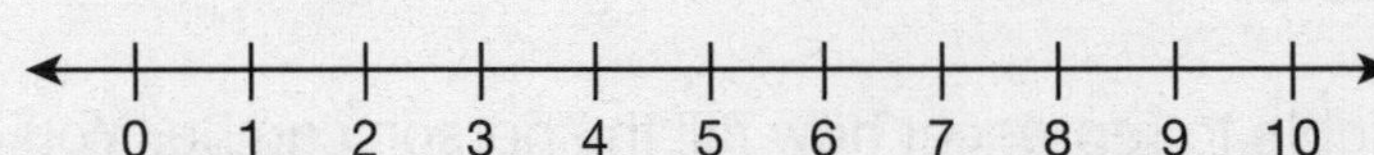

4. A bag must weigh less than 16 pounds to be carried on an airplane. Let p = the number of pounds the bag must weigh.

   Inequality: ____________ Possible solution: ____________

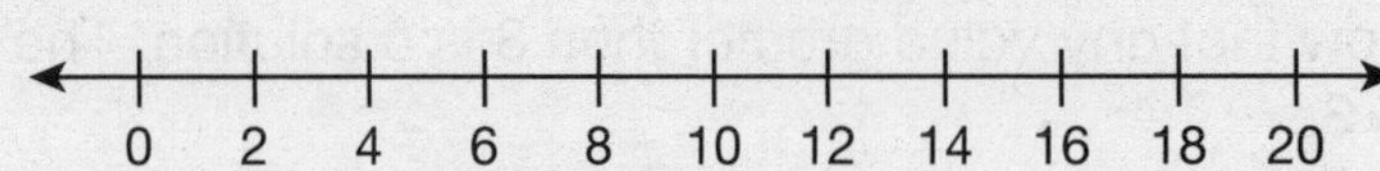

   Explain how you wrote and graphed the inequality and why you chose your solution.

   ____

   ____

CCSS: 6.EE.9

Lesson 22: Applying Equations

Some equations have two variables where the value of one depends on the value of the other. In this type of equation, the variable that depends on the other is called the **dependent variable**. The **independent variable** does not depend on the other variable.

Example

Maureen reads 2 pages every minute. To represent this rate, she writes the equation $p = 2m$, where p is the number of pages she reads and m is the number of minutes. Which are the independent and dependent variables?

The equation is $p = 2m$. The number of pages Maureen reads depends on how many minutes she spends reading. The number of minutes she reads is up to Maureen. It is not dependent on another variable.

Therefore, p, the number of pages read, is the dependent variable, and m, the number of minutes Maureen spends reading, is the independent variable.

Example

Taro earns a salary of $8.75 per hour. To determine how much money he makes each week, he uses the equation $s = 8.75h$, where s is his total weekly salary and h is the number of hours he works in a week. Which are the independent and dependent variables?

The equation is $s = 8.75h$. The total amount of money Jackson makes is dependent on the number of hours he works. The number of hours that Jackson works can be changed and affects how much money he makes.

Therefore, s, Taro's weekly salary, is the dependent variable, and h, the number of hours Taro works in a week, is the independent variable.

A table can show the relationship between a dependent and an independent variable.

Example

For every minute that Maxine runs, she travels a distance of 700 feet. Write an equation with two variables to show the total distance run. Identify the dependent and independent variables. Then create a table of values.

For each minute, the total distance is equal to 700 feet. The total distance is equal to the number of minutes multiplied by 700. Use *m* for the number of minutes that Maxine runs. Use *d* for the total distance she travels. The following equation shows this relationship. $d = 700m$

The total distance is dependent on the number of minutes Maxine runs. Therefore, *d* is the dependent variable. The number of minutes that Maxine runs can be changed by Maxine. The independent variable is *m*.

The following table shows the relationship between the variables.

Number of Minutes run, *m*	Total Distance in Feet, *d*
1	700
2	1400
3	2100
4	2800

Example

For every hour that Hakeem talks to Europe on his cell phone, he is charged $6. Write an equation with two variables to show Hakeem's total charge. Identify the dependent and independent variables. Then create a table of values.

For each hour, the charge is equal to $6. The total charge, therefore, should be equal to the number of hours multiplied by $6. Use *h* for the number of hours that Hakeem talks to Europe. Use *c* for the total charge. The following equation shows this relationship. $c = 6h$

The total charge is dependent on the number of hours that Hakeem talks to Europe on his phone. Therefore, *c* is the dependent variable. The number of hours that Hakeem talks on the phone is up to him and is not dependent on the total charge. The independent variable is, therefore, *h*.

You can construct a table that shows the relationship between the two variables. Pick different values for the independent variable, and find the corresponding value of the dependent variable using the equation.

Number of Hours on phone, *h*	Total Charge in dollars, *c*
1	6
2	12
3	18
4	24
5	30

Practice

Directions: For questions 1 through 6, identify the independent and dependent variables.

1. Mr. Edgington travels 25 miles per hour in his boat. To determine how far he travels, he uses the equation $d = 25h$, where d is the total distance in miles and h is the number of hours he travels in his boat.

   independent variable: ______________ dependent variable: ______________

2. There are 4 energy-efficient light bulbs in each pack. The equation $b = 4p$ shows how many light bulbs in all, b, are in p number of packs of light bulbs.

   independent variable: ______________ dependent variable: ______________

3. A satellite travels 7 km over Earth during each second of its orbit. To calculate the satellite's total distance, a scientist uses the equation $d = 7s$, where d is the total distance in km and s is the number of seconds of orbit.

   independent variable: ______________ dependent variable: ______________

4. Each train car has a capacity of 68 passengers. To determine the capacity of a train, an engineer uses the equation $t = 68c$, where t is the total capacity and c is the number of cars on the train.

   independent variable: ______________ dependent variable: ______________

5. An organic grocer charges $2.50 for each avocado. The cost of a number of avocados is found with the equation $c = 2.50a$, where c is the total cost and a is the number of avocados purchased.

   independent variable: ______________ dependent variable: ______________

6. Every time Jill puts a dime into a parking meter, she gets 15 minutes to park her car on the street. She uses the equation $t = 15d$, where t is the total amount of time in minutes she can park her car and d is the number of dimes she puts into the meter.

   independent variable: ______________ dependent variable: ______________

Directions: For questions 7 and 8, write an equation based on the real-world situation. Then use the equation to fill in the table of values.

7. A young elm tree grows 3 feet every year. Write an equation to show the total height of an elm tree, *t*, based on *y* years.

____

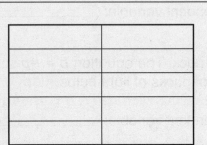

8. Each floor of a skyscraper is 11 feet tall. Write an equation to show the height of a skyscraper in feet, *h*, based on the number of floors, *f*.

____

9. For each new ink cartridge, a printer can print 250 pages. Write an equation to show the total number of pages, *p,* that can be printed from *c* ink cartridges. Then fill in the following table.

____

Explain how your table represents the relationship between the variables.

____

____

____

Unit 3 Practice Test

Read each question. Choose the correct answer.

1. What is the value of x in the following equation?

$$4x = 6$$

A. $\frac{2}{3}$

B. $1\frac{1}{2}$

C. 2

D. 24

2. What mathematical term can be used to describe the underlined part of the following expression?

$$\underline{7}a - 3$$

A. coefficient

B. operation

C. term

D. variable

3. Which expression is equivalent to the following expression?

$$2x + 12 + 3x + y + 13$$

A. $6x + 25$

B. $31x$

C. $5x + 25y$

D. $5x + y + 25$

4. What is the value of the following expression if $n = 3$?

$$20 - n^2 \times 2$$

A. 2

B. 8

C. 22

D. 28

5. Which inequality represents the following scenario if b is the number of pages of Carmen's book report?

Carmen's book report needs to be more than 4 pages long.

A. $4 > b$

B. $b < 4$

C. $b = 4$

D. $b > 4$

6. What is the value of the following expression?

$$(3 + 2)^2 - 3 \times 2$$

A. 1

B. 16

C. 19

D. 44

7. Write which variable is dependent and which one is independent in the following scenario and equation.

   For every gallon of gas that Eliza puts in her car, she can drive 35 miles. She determines how far she can drive using the equation $d = 35g$, where d is the total distance, in miles, and g is the number of gallons she puts in her car.

   d: ______________________________

   g: ______________________________

8. Alejandra needs to bring more than 6 kg of trail mix on a camping trip. She writes the following inequality to represent how much trail mix she must bring.

   $t > 6$, where t = the weight of the trail mix in kg

   Graph Alejandra's inequality on the following number line.

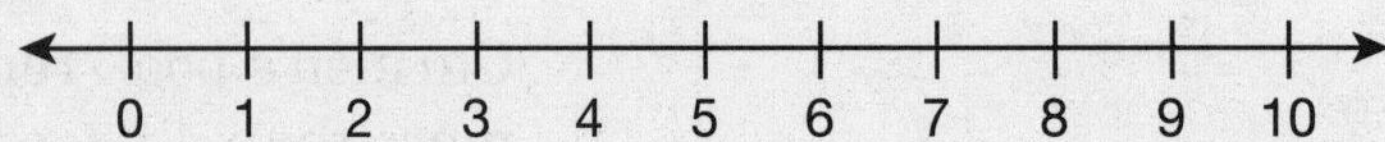

9. What is the value of $6 + 4 \times 5 - 1$?

   ____

10. Brendan hit 4 more home runs this year than he did last year. Write an expression to show how many home runs he hit this year. Let h = the number of home runs he hit last year.

    ____

11. Write an expression that is equivalent to $4(x + 7y)$.

    ____

12. Are the following two expressions equivalent? Write *yes* or *no.*

    $4n + 2n + 2 + 2n$
    $2n + 4 + 2n + 2n$

    ____

13. What is the solution to $3x < 18$?

____

14. Write an algebraic expression to represent the following expression.

Thirteen less than g

____

15. Emily determined that the solution to the following equation is $z = 3$.

$$3z = 3$$

Use substitution to determine if Emily was correct. Write *correct* or *incorrect*.

____

16. Every apple that Frankie buys costs \$1.29. He uses the equation $c = 1.29a$ to determine the total cost, c, of buying a apples. Is the variable a the dependent or independent variable?

____

17. What is the value of the following expression if $x = 8$?

$$10 + x \div 2 - 1$$

____

18. Ira has less than 40 minutes to finish his homework. Write an inequality to represent how much time Ira can use to finish his homework. Let m = the number of minutes Ira will use to finish his homework.

____

19. Using the same terms, write an expression that is equivalent to $(3 + 7) + 4$.

____

20. Solve the following equation for q.

$$q + 16 = 25$$

____

21. What are the two factors in the following expression?

$$2a(3 + 6)$$

__________ and __________

22. Write an expression that is equivalent to the following expression.

$$b + b + c + c + 2b$$

____

23. Suman determined that the solution to the following equation is $k = 18$.

$$13 + k = 31$$

Use substitution to determine if Suman was correct. Write *correct* or *incorrect*.

____

24. What is the value of $25 - 4 \times 3 + 1$?

____

25. Solve the following equation for r.

$$r + 0.25 = 3.5$$

____

26. Write an algebraic expression to represent the following expression.

the product of forty and m

____

27. Every minute that Maria swims, she burns 8 calories. She uses the equation $c = 8m$ to determine the total calories burned, c, for swimming m minutes. Is c the dependent variable or independent variable?

____

28. What is the coefficient in the following expression?

$$20x - 5$$

____

29. Write an expression that is equivalent to the following expression.

$$2(9w + x)$$

____

30. The formula to convert temperature from Fahrenheit to Celsius is $C = \frac{5}{9}(F - 32)$. What is the Celsius temperature, C, if the Fahrenheit temperature, F, is 59°?

____

31. Solve the following equation for p.

$$5p = 55$$

____

32. The cargo on an airplane must weigh less than 12 tons. The pilot uses the following inequality to represent the possible weight of the plane's cargo.

$c < 12$, where c = the weight of the cargo in tons

Graph the pilot's inequality on the following number line.

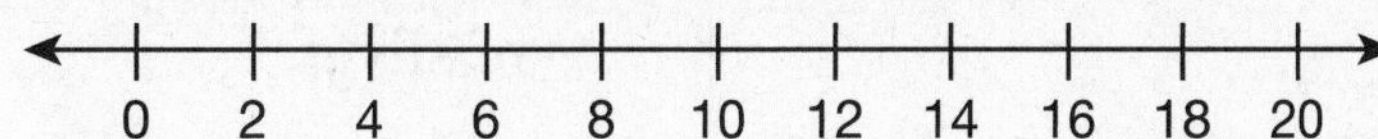

33. Each week, Tino earns $35 by mowing his neighbors' yards. He wants to find out how much money he can make if he works different numbers of weeks. He uses the equation $m = 35w$, where m is the total amount of money he earns, in dollars, and w is the number of weeks he mows his neighbors' lawns. Complete the following table.

Number of weeks, *w*	Total money earned, *m* (in dollars)

34. Are the following two expressions equivalent? Write *yes* or *no.*

$$10 + b + 2 + 9b \qquad 2(5b + 6)$$

____

35. Write which variable is dependent and which one is independent in the following scenario.

For each hour that a ceiling fan is on, the fan uses 75 watts of energy. To determine how much energy it uses, you can use the equation $e = 75h$, where e is the total energy used, in watt-hours, and h is the number of hours that the ceiling fan is on.

e: _______________

h: _______________

36. What is the value of the following expression?

$$6(3 - 1) \div 4 + 2$$

____

37. What is the solution of $x - 15 \geq -19$?

____

38. Matthew determined that the solution to the following equation is $y = 2\frac{1}{2}$.

$$4y = 10$$

Use substitution to determine if Matthew was correct. Write *correct* or *incorrect*.

____

39. Every song that Taryn buys online costs $0.99. She uses the equation $c = 0.99s$ to determine the total cost, c, of buying s songs. Is c the dependent variable or independent variable in her equation?

____

40. The area of a square is $A = s^2$, where A is the area and s is the length of a side of the square. What is the area of a square where a side length $s = 50$ feet?

____

41. A jeweler needs more than 3.2 ounces of silver to make a pendant. Write an inequality to represent how much silver the jeweler needs to make the pendant. Let s = the number of ounces of silver.

____

42. Write an expression that is equivalent to $5 \times (6 \times 7)$. Use the same terms in your expression.

____

43. Solve the following equation for t.

$$\frac{3}{8} + t = \frac{7}{8}$$

____

44. What is the variable in the following expression?

$$9(9 + 2z)$$

____

45. Write an expression that is equivalent to the following expression.

$$3(4 + 3x) + y + x$$

____

46. The total weight of the freight in an industrial elevator must be less than 3 tons.

Part A

Write an inequality to show the possible total weight of the freight in the elevator. Let w = the total weight, in tons.

____

Part B

Graph the inequality on the following number line.

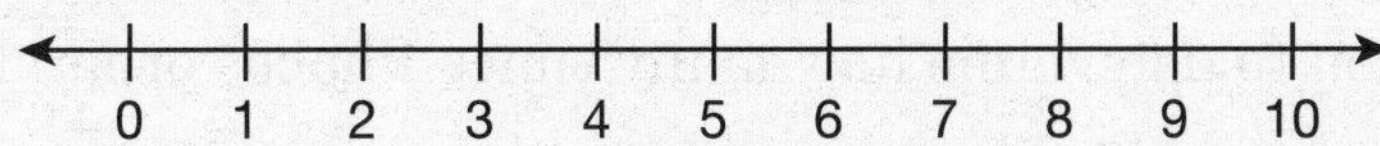

____

Part C

Write three different solutions for the inequality.

____

____

____

Explain why you knew that the three numbers you wrote were solutions to the inequality.

____

____

47. Phobos, the larger of Mars's two moons, orbits Mars three times each day.

Part A
Write an equation to show the relationship between the number of days and the number of times Phobos orbits Mars. Use d for the number of days and p for the number of times Phobos orbits Mars.

____

Part B
Use your equation to determine how many times Phobos orbits Mars in 9 days.

____

Part C
Complete the following table to show how the variables in the equation are related.

Number of days, *d*	Number of Times Phobos Orbits Mars, *p*

Explain how you determined the numbers in the right-hand column based on the numbers in the left-hand column.

____

____

____

Geometry

Geometry can be applied to almost everything you see in the world around you. After all, the word *geometry* literally means "measure of the Earth." The natural world is filled with geometric shapes, so it is often described and measured in terms of geometry. For example, you can measure the size of your bedroom by finding its area. A mapmaker might draw a map of your town on a coordinate plane so that you can locate your school—or any building in town—with an ordered pair. Much of the human-made world is constructed according to geometric principles.

In this unit, you will review how to find the areas of two-dimensional figures. You will find the volume of three-dimensional figures. You will use nets to see how three-dimensional figures are represented in two-dimensions. Finally, you will use the coordinate plane to create polygons and determine the distance between their vertices.

In This Unit

Lesson 23: Area of Figures

Area (*A*) is the measure of the region inside a two-dimensional figure. Area is measured in square units (units^2). You can find the area of a figure by counting the squares inside it. You can also find the area of a rectangle by multiplying its length by its width.

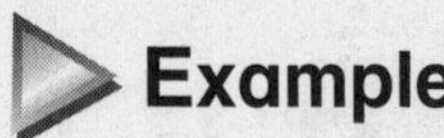

Example

This rectangle has a length of 6 in. and a width of 3 in. What is its area?

First, break up the length and width into 1-in. pieces. There are six 1-in. pieces in the length. There are three 1-in. pieces in the width. In total, there are eighteen 1-in. squares inside the rectangle, so the area of the rectangle is 18 in.^{2}. However, you can also find the area by multiplying its length by its width.

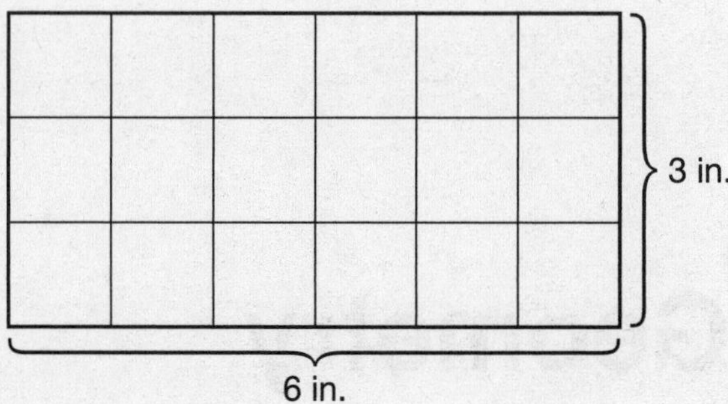

The length of the rectangle is 6 in. The width is 3 in. The area of the rectangle is equal to 6 in. $\times$ 3 in., which is 18 in.^{2}. The product matches the number of squares inside the rectangle. The area is 18 square inches or 18 in.^{2}.

The area of a triangle is one-half the area of a rectangle with the same height and length.

Example

This right triangle has a length of 5 ft and a height of 4 ft. What is its area?

The triangle cuts the 1-ft squares into irregular pieces, so you cannot count the squares inside the figure. However, the triangle has half the area of the rectangle. Therefore, to find the area of a triangle, you can divide the area of the rectangle by 2.

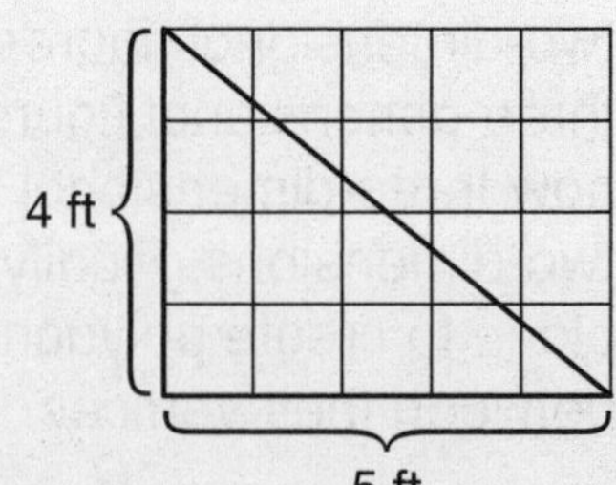

The area of the rectangle is the length times the width: 5 ft $\times$ 4 ft $=$ 20 ft^2. Divide by 2 to find the area of the triangle: 20 ft^2 $\div$ 2 $=$ 10 ft^2. The area of the right triangle is 10 square units.

CCSS: 6.G.1

It does not matter whether a triangle is a right triangle. The area of any triangle is one-half the area of a rectangle with the same height and length.

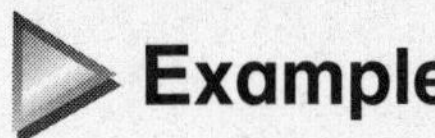

Example

What is the area of the following obtuse triangle?

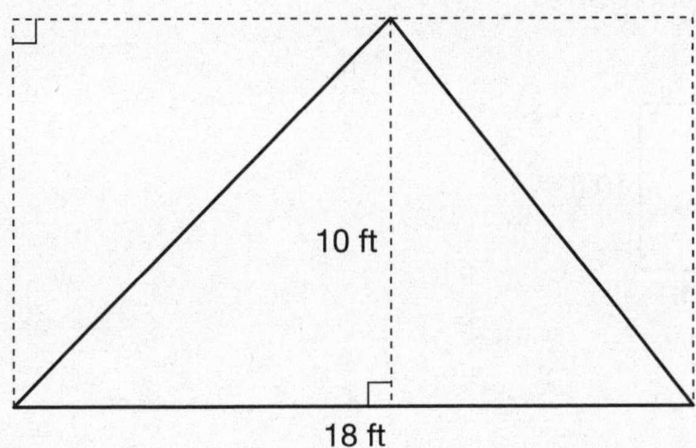

The area of the triangle is equal to half the area of a rectangle with the same length and height. The area of a rectangle with a length of 18 ft and a height of 10 ft would be 18 ft $\times$ 10 ft $=$ 180 ft^2. The area of the triangle is equal to 180 ft^2 $\div$ 2, or 90 ft^2.

To find the area of a parallelogram, multiply the length by the height. The height is the distance from the bottom of the figure to the top.

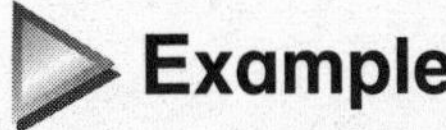

Example

What is the area of the following parallelogram?

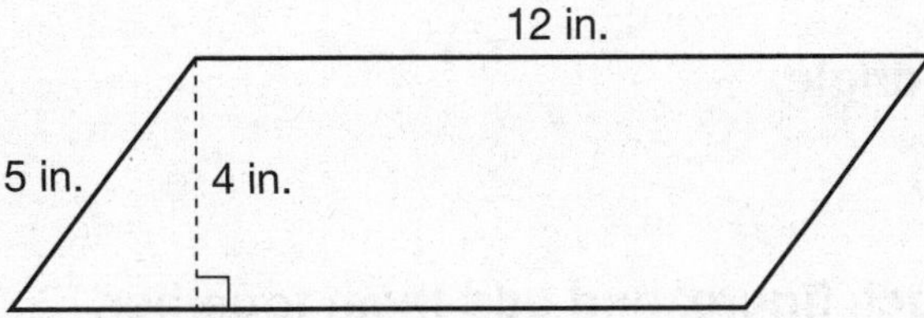

Remember that the *height and length* of a parallelogram are important—not the length of the sides. To find the area of the parallelogram, multiply the length by the height.

The length of the parallelogram is 12 in. Its height is 4 in. The area of the parallelogram is 12 in. $\times$ 4 in., or 48 in.^{2}.

To find the area of an irregular polygon, you can often break it into triangles and rectangles. Then find the area of each part. Finally, add the areas together to find the total area of the polygon.

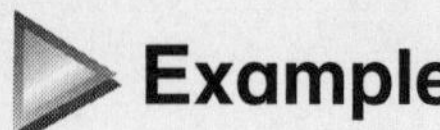

Example

What is the area of this polygon?

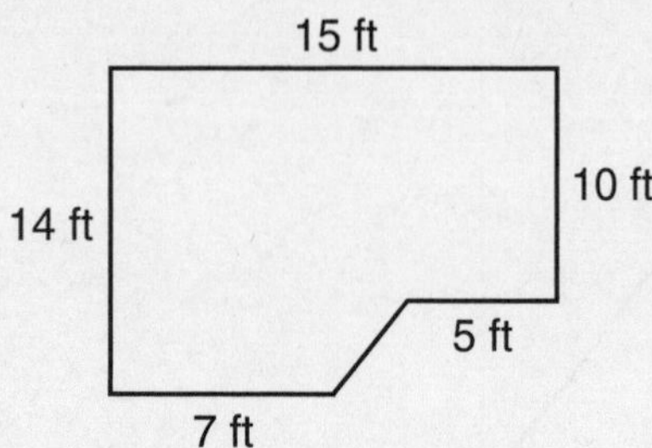

Step 1: **Divide the polygon into rectangles and/or triangles.**

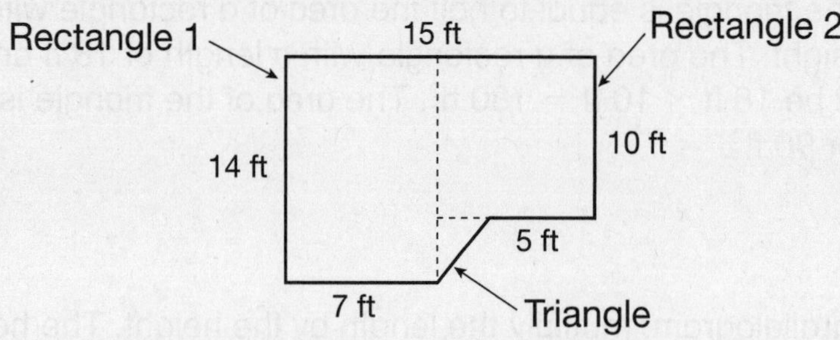

You can divide the polygon into two rectangles and a triangle.

Step 2: **Find the height of the triangle.**

14 ft − 10 ft = 4 ft

Step 3: **Find the base of the triangle.**

15 ft − (7 ft + 5 ft) = 3 ft

Step 4: **Now find the area of each figure, and add them together.**

Rectangle 1: 7 ft × 14 ft = 98 ft^2

Rectangle 2: 10 ft × (5 ft + 3 ft) = 80 ft^2

Triangle: $\frac{1}{2}$(3 ft × 4 ft) = 6 ft^2

The total area is 98 ft^2 + 80 ft^2 + 6 ft^2 = 184 ft^2.

There are many situations where you'll need to find the area of a figure or shape.

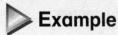

Example

Patricia has a rectangular garden with a length of 14 m and a width of 6 m. Patricia plants petunias in the shaded area of the garden, shown below. How much area of her garden is covered by petunias?

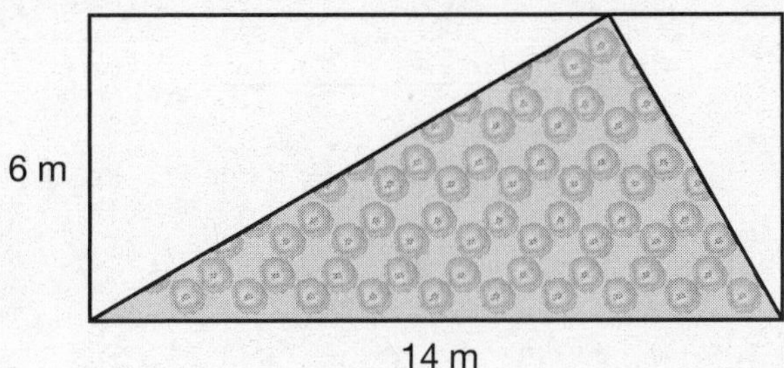

The area of Patricia's garden is equal to its length times its height. That is 14 m $\times$ 6 m, or 84 m^2. The triangular part of the garden is equal to half that area: 84 m^2 $\div$ 2, or 42 m^2.

The area of Patricia's garden covered by petunias is 42 m^2.

TIP: When a triangle is inscribed within a rectangle, the area of the triangle is equal to the area of the other parts outside the triangle. In the example above, the area of Patricia's petunias is half the area of the total rectangle.

Practice

Directions: For questions 1 through 5, find the area of each triangle or rectangle.

1. 

$A =$ ____________

2. 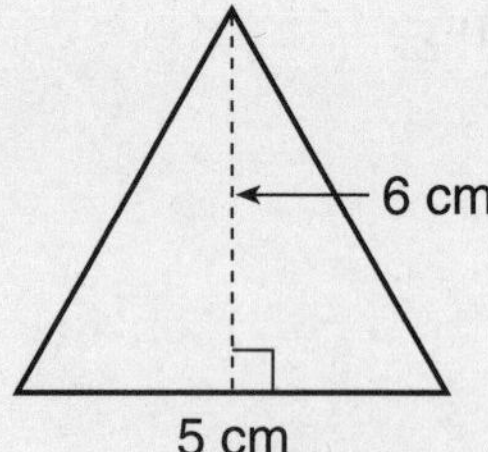

$A =$ ____________

3. 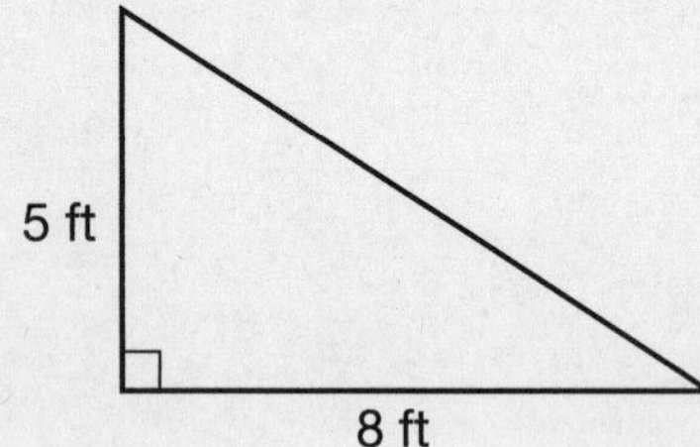

$A =$ ____________

4. 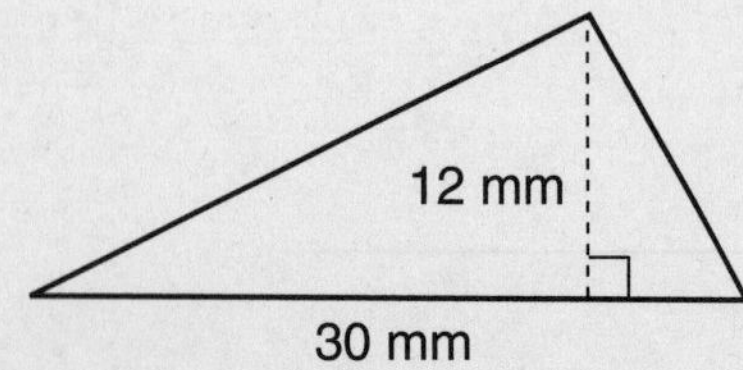

$A =$ ____________

5.

$A =$ ____________

Directions: For questions 6 through 9, find the area of each quadrilateral.

6.

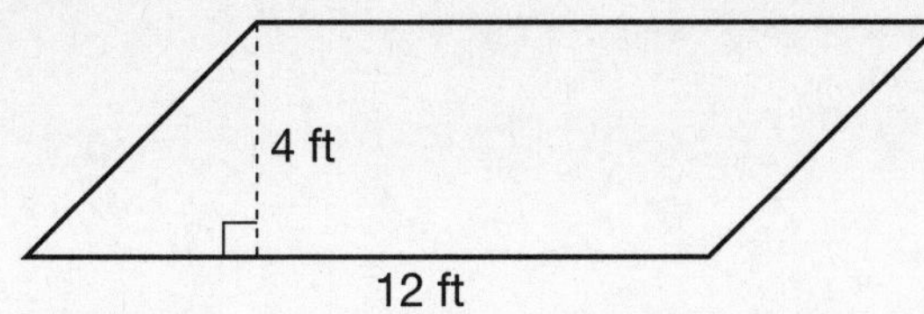

$A =$ ______________

7.

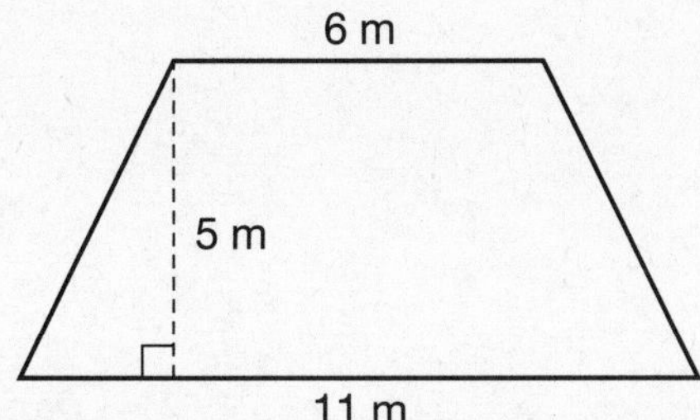

$A =$ ______________

8.

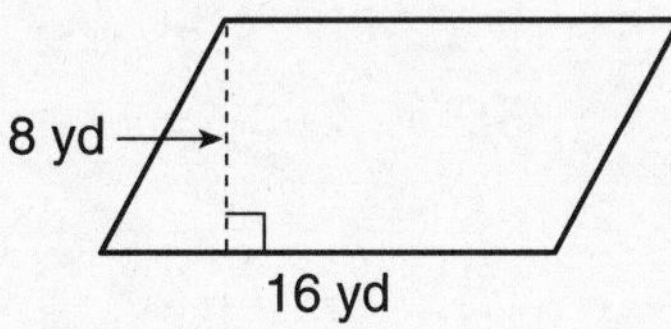

$A =$ ______________

9. What is the area of the following trapezoid?

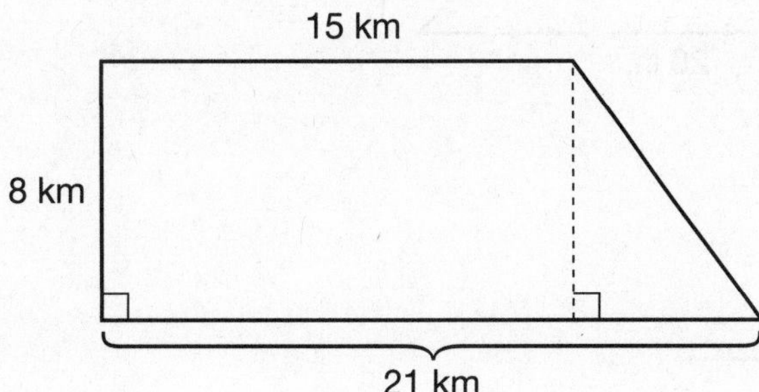

A. 120 km^2
B. 144 km^2
C. 148 km^2
D. 168 km^2

Directions: For questions 10 through 13, find the area of each polygon.

10.

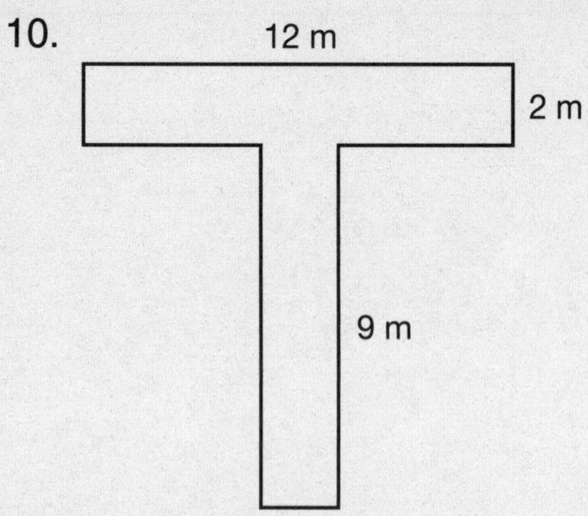

$A =$ ______________

11.

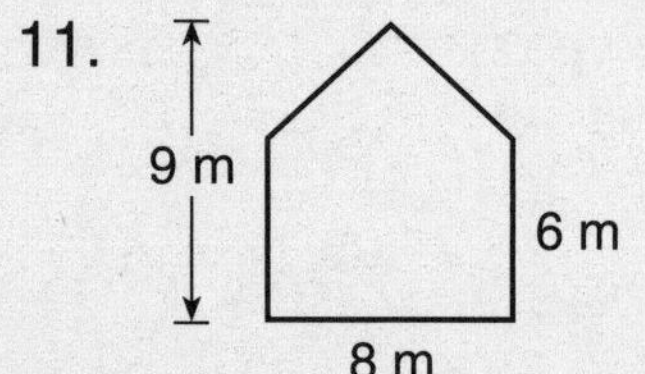

$A =$ ______________

12.

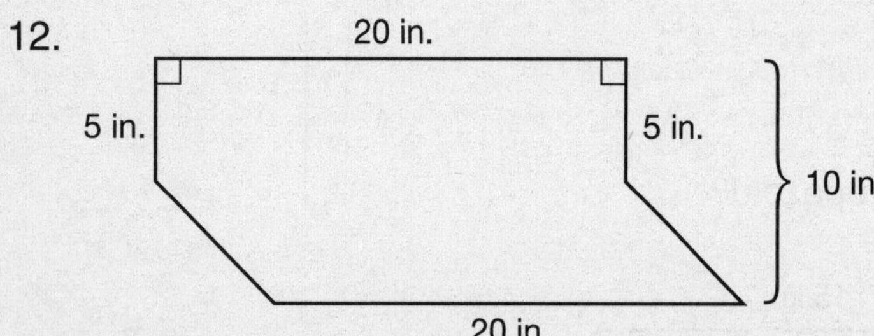

$A =$ ______________

13.

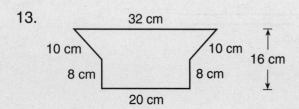

32 cm
10 cm
10 cm
16 cm
8 cm
8 cm
20 cm

$A =$ ______________

CCSS: 6.G.1

14. Hitomi's bedroom is represented by the figure shown below.

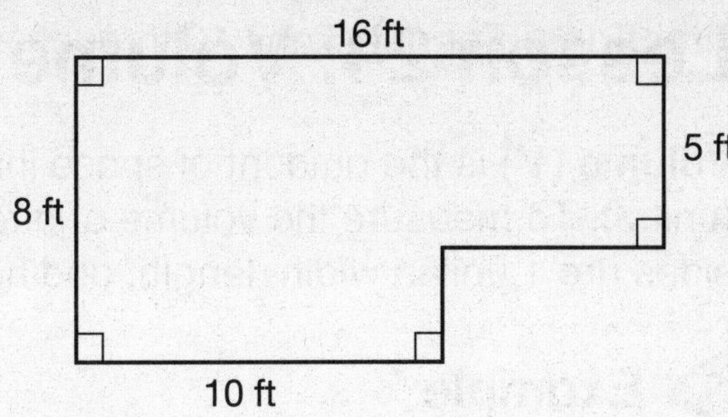

What is the area of Hitomi's bedroom, ______________?

15. The figure shows the shape of a city park.

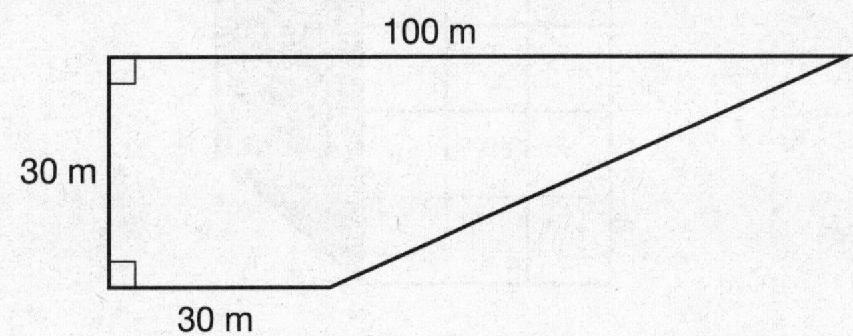

A landscaper needs to plant grass seeds in the park. To determine how many seeds to use, she needs to know the area of the park. What is the area of the park?

____

16. A shipping company makes different-sized sails for sailboats. The following figure shows the shapes of two sails used on one sailboat.

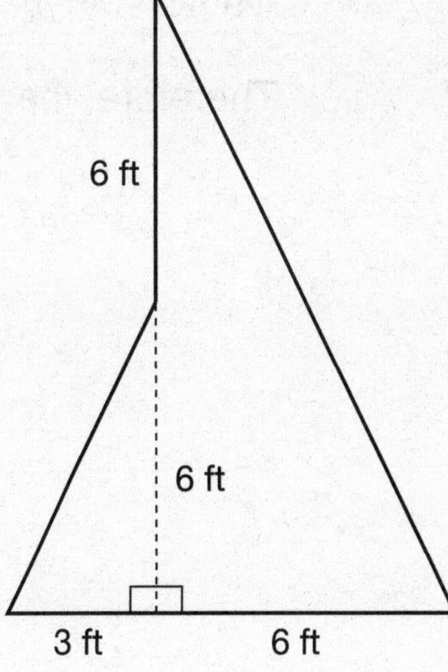

What is the combined area of the two sails? ______________________

Explain how you found your answer.

____

Lesson 24: Volume of Rectangular Prisms

Volume (*V*) is the amount of space inside a solid. Volume is measured in cubic units (units^3). To measure the volume of a rectangular prism, you can fill it with cubes whose sides are 1 unit in width, length, and height.

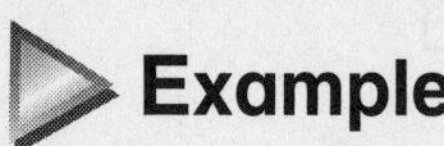

Example

Determine the volume of the rectangular prism below.

The prism is made up of a collection of cubes. Each cube has a width, length, and height of 1 unit. Therefore, the volume of the prism will be equal to the number of unit cubes that compose it.

To find the number of cubes, you need to determine the length, width, and height of the prism. The prism is 3 cubes long, 4 cubes wide, and 4 cubes high. So, there are 12 cubes on each of the 4 levels of the prism.

Therefore, the volume is $12 \times 4 = 48$ cubic units or 48 units^3.

CCSS: 6.G.2

You can find the volume of a rectangular prism with fractional edge lengths.

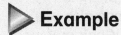

Example

A rectangular prism has a length of $\frac{6}{8}$ in., a width of $\frac{4}{8}$ in., and a height of $\frac{4}{8}$ in. What is the volume of the rectangular prism?

In the illustration below, the rectangular prism is drawn so that it is made up cubes that are $\frac{1}{8}$ in. in length, width, and height. Each cube has a volume of $\frac{1}{512}$ in.^{3}. Because the length of the prism is $\frac{6}{8}$ in., the prism is 6 cubes long. Because its width is $\frac{4}{8}$ in., the prism is 4 cubes wide. Because its height is $\frac{4}{8}$ in., the prism is 4 cubes tall.

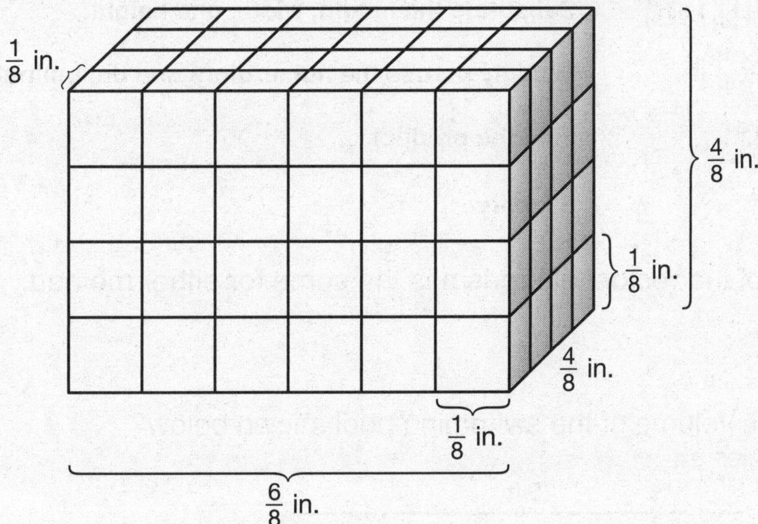

You can now use the model to find the volume of the prism. The prism has 6 by 4 by 4 cubes. Each of the 4 rows has 24 cubes, a total of ninety-six $\frac{1}{512}$ in.^{3} cubes. Multiply the number of cubes by the volume of each cube to find the total volume:

$$V = 96 \times \frac{1}{512}$$
$$= \frac{96}{512}$$
$$= \frac{3}{16}$$

Therefore, the volume is equal to $\frac{3}{16}$ in.^{3}.

You can also determine the volume of a rectangular prism by multiplying the edge lengths of the prism. The formula for finding the volume of a rectangular prism is $V = lwh$, where V is volume, l is length, w is width, and h is height. You can also use $V = Bh$, where B is the area of the base of the prism and h is its height. Whether you use a diagram showing unit cubes or one of these formulas, the product will be the same.

Example

Use the example of the rectangular prism from the previous page. Find the volume using a formula.

$V = lwh$

$V = \left(\frac{6}{8}\right)\left(\frac{4}{8}\right)\left(\frac{4}{8}\right)$ in.^{3} **Substitute the length, width, and height.**

$V = \frac{6 \times 4 \times 4}{8 \times 8 \times 8}$ in.^{3} **Multiply across the numerators and denominators.**

$V = \frac{96}{512}$ in.^{3} **Find the product.**

$V = \frac{3}{16}$ in.^{3} **Simplify.**

The area of the rectangular prism is the same for either method.

Example

What is the volume of the swimming pool shown below?

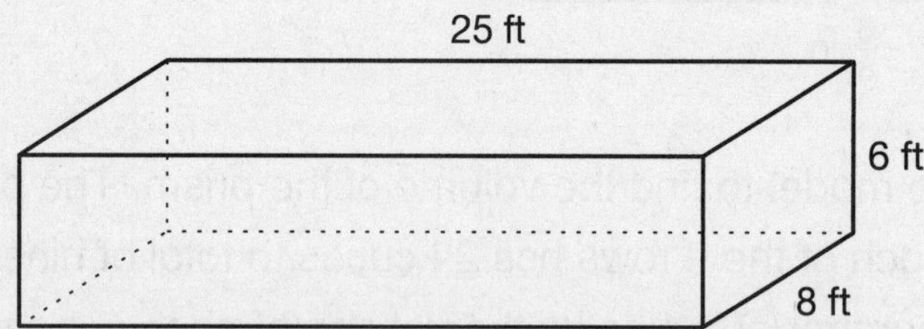

The area of the base of a rectangular prism is the product of the length and the width. The base is 25 ft × 8 ft = 200 ft^2. Now use the formula $V = Bh$.

$V = Bh$

$V = (200\text{ ft}^2)(6\text{ ft})$ **Substitute the base and height.**

$V = 1200\text{ ft}^3$ **Multiply.**

The volume of the swimming pool is 1200 ft^3.

Practice

Directions: For questions 1 through 3, find the volume of the rectangular prism.

1.

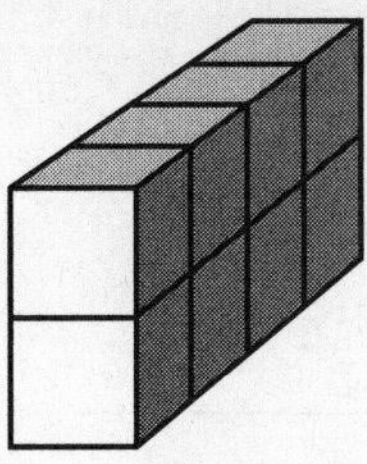

$V =$ ______________

2.

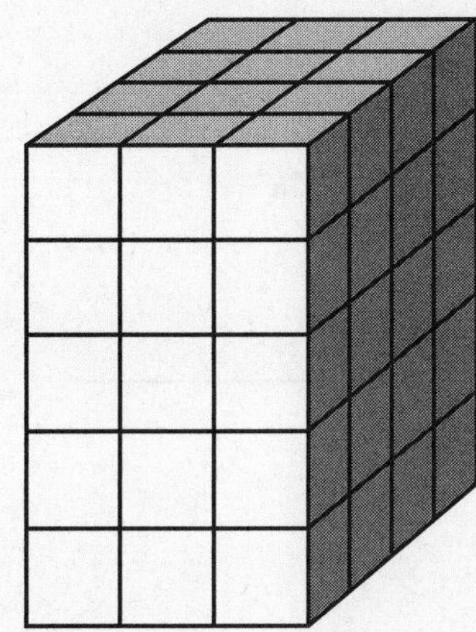

$V =$ ______________

3.

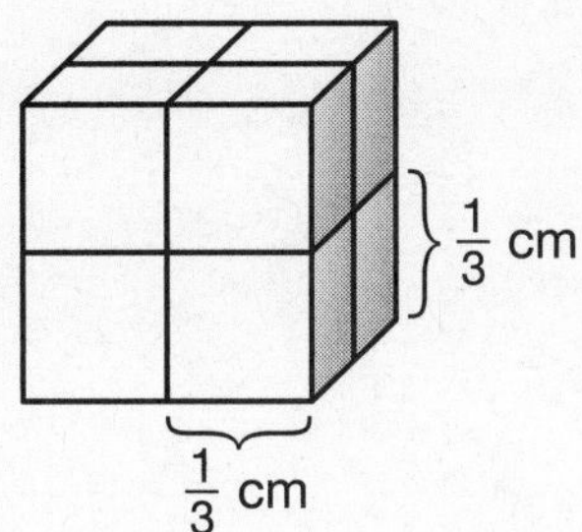

$V =$ ______________

Directions: For questions 4 through 7, find the volume for each rectangular prism.

4.

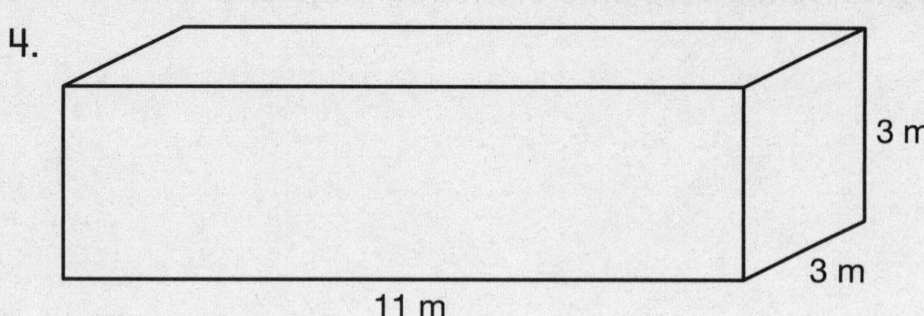

$V =$ ______________

5.

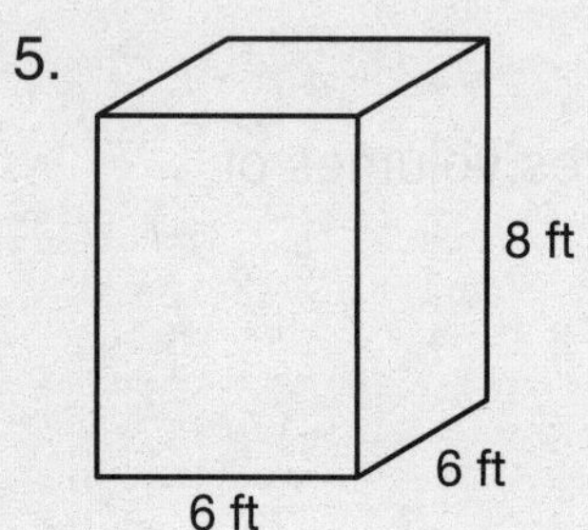

$V =$ ______________

6.

$V =$ ______________

7.

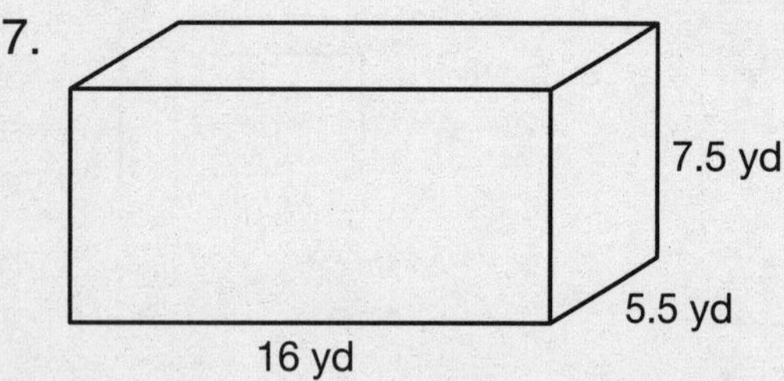

$V =$ ______________

8. A cereal company wants to determine how much cereal can fit in its cereal box, shown below.

What is the volume of the cereal box? ______________

9. Laura wants to buy a large fridge for her kitchen. She compares volumes of different fridges, including the model shown below.

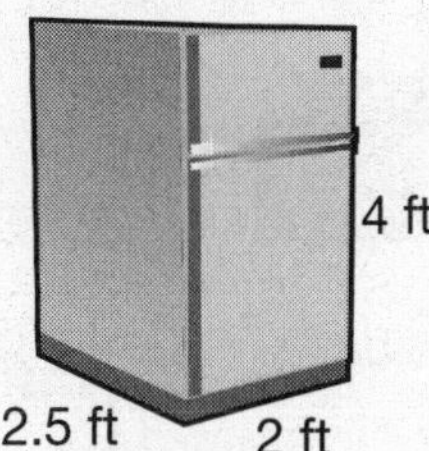

What is the volume of the refrigerator? ______________

10. Hobart needs to mail a package for work. He wants to use the following box, but he is not sure if the box is big enough.

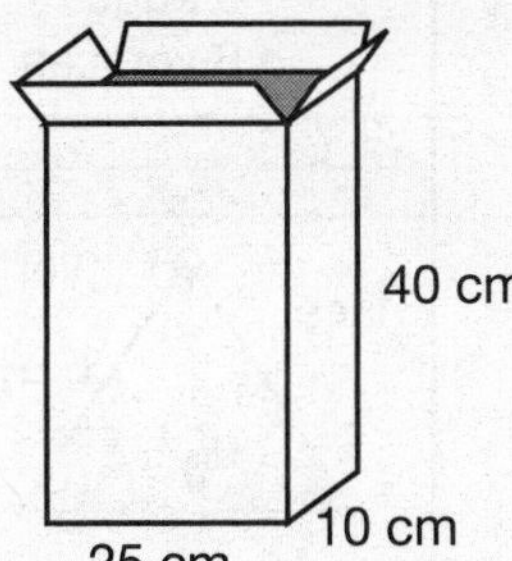

What is the volume of Hobart's box? ______________

Explain how you found your answer.

____

____

Lesson 25: Nets

A **net** is a two-dimensional representation of a solid. A net shows all the faces, edges, and vertices of a solid.

The **faces** are the plane figures that make up the sides and the base(s) of the solid. The faces intersect to form the **edges** of the figure. The point of intersection of three or more edges is a **vertex**, or corner of the figure.

The following table shows some solid figures and their nets. Prisms and pyramids can have any polygon as a base. A cube is a rectangular prism that has all square faces. A pyramid with a square base is called a **square pyramid** and has many of the same properties as a rectangular pyramid.

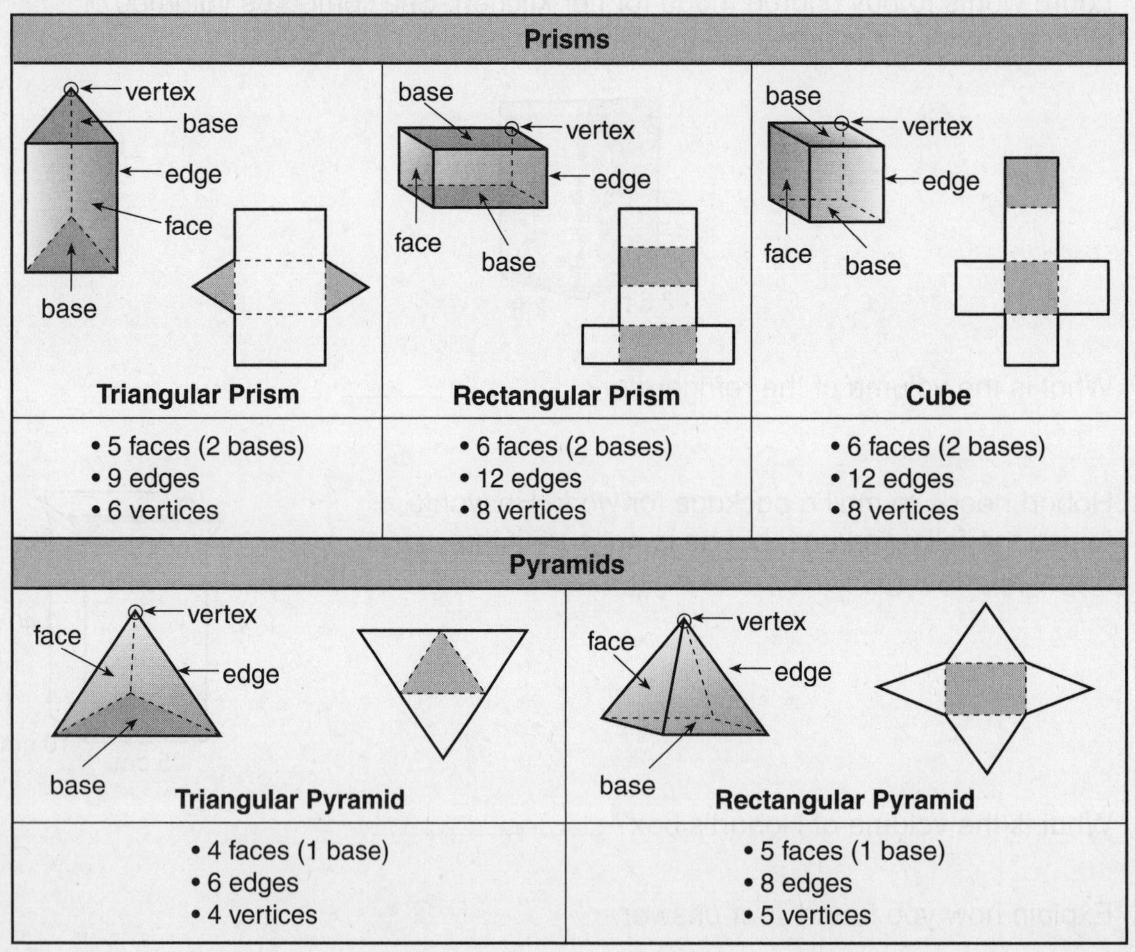

Prisms		
Triangular Prism	Rectangular Prism	Cube
• 5 faces (2 bases) • 9 edges • 6 vertices	• 6 faces (2 bases) • 12 edges • 8 vertices	• 6 faces (2 bases) • 12 edges • 8 vertices

Pyramids	
Triangular Pyramid	Rectangular Pyramid
• 4 faces (1 base) • 6 edges • 4 vertices	• 5 faces (1 base) • 8 edges • 5 vertices

CCSS: 6.G.4

To create a net from a solid, consider how many faces the solid has. That will be the number of shapes in the net. Consider the shapes that are used to make the faces of the solid. Those will be the shapes in the net. The shapes in the net must be able to fold together to form the solid figure.

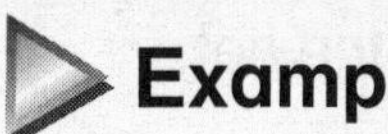

Example

Create a net from the following figure.

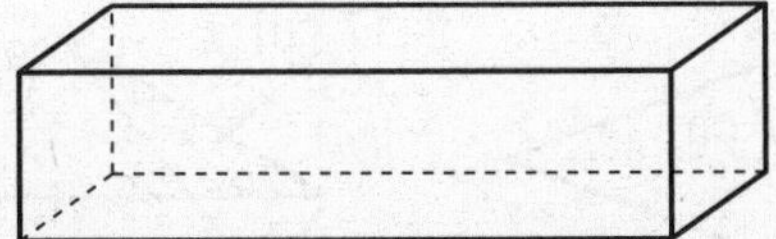

The figure is a rectangular prism. It has 6 rectangular sides. Of the 6 rectangular sides, 2 are squares. The net must be made up of 4 rectangles and 2 squares.

The front face of the prism is a long rectangle with the square faces attached to its ends. The following image shows those three faces of the prism.

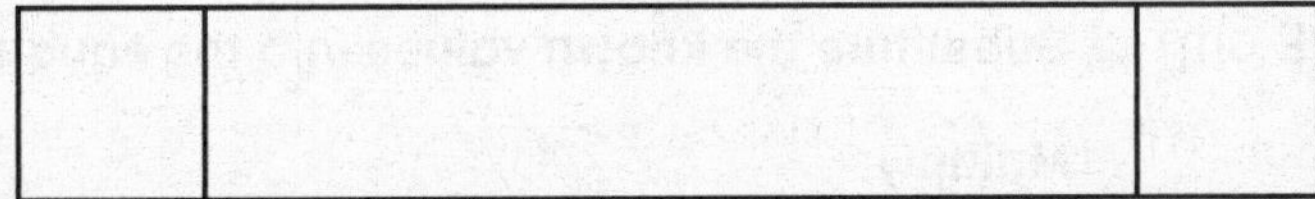

There are three other long rectangles that make up the top, back, and bottom of the prism. The following figure is a possible net for the prism.

TIP: There may be several possible nets for a solid figure. In the example on this page, the square faces could be shown to the left and right of any of the rectangular faces. The resulting nets could still be used to construct the given solid figure.

Surface area (S.A.) is the measure of the outside area of a three-dimensional figure. To find the surface area, find the sum of the areas for all of the faces in the solid's net.

Example

Find the surface area of the rectangular pyramid represented by the net.

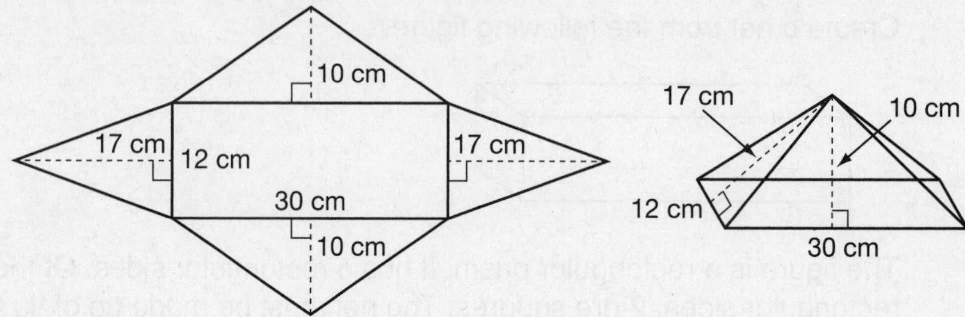

The solid figure has 5 faces. To find its surface area, determine the area of each face. Begin with the one rectangular face:

$A = lw$

$A = (30\text{ cm})(12\text{ cm})$ **Substitute the known values into the equation.**

$A = 360\text{ cm}^2$ **Multiply.**

Now determine the areas of the top, right, bottom, and left triangles. Use the formula for the area of a triangle. Substitute for the known values and simplify.

Top triangle

$A = \frac{1}{2}bh$

$A = \frac{1}{2}(30\text{ cm})(10\text{ cm})$

$A = 150\text{ cm}^2$

Bottom triangle

$A = \frac{1}{2}bh$

$A = \frac{1}{2}(30\text{ cm})(10\text{ cm})$

$A = 150\text{ cm}^2$

Right triangle

$A = \frac{1}{2}bh$

$A = \frac{1}{2}(12\text{ cm})(17\text{ cm})$

$A = 102\text{ cm}^2$

Left triangle

$A = \frac{1}{2}bh$

$A = \frac{1}{2}(12\text{ cm})(17\text{ cm})$

$A = 102\text{ cm}^2$

The surface area is the sum of the areas of all the faces.

$360\text{ cm}^2 + 150\text{ cm}^2 + 150\text{ cm}^2 + 102\text{ cm}^2 + 102\text{ cm}^2 = 864\text{ cm}^2$

The surface area of the rectangular pyramid is 864 cm^2.

CCSS: 6.G.4

Example

A sculptor designs an outdoor sculpture in the shape of a cube that has a side length of 8 feet. The following diagram shows the cube and its net.

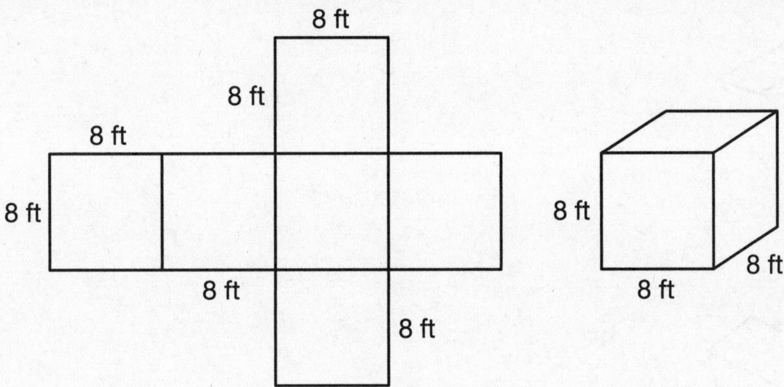

The city needs to paint the sculpture every year. To determine how much paint is needed, it needs to find the surface area of the cube. What is the surface area of the sculpture?

The solid figure has 6 faces. To find its surface area, determine the area of each face. Because the solid is a cube, each face is a square. Use the formula for the area of a square.

$A = s^2$

$A = (8)^2$ **Substitute the known values into the equation.**

$A = 64$ **Multiply.**

Now you can multiply the area of one face of the cube by 6, the number of faces, to find the surface area of the cube.

$64 \times 6 = 384$

The surface area of the cube-shaped sculpture is 384 ft^2.

Practice

Directions: For questions 1 through 4, draw the net of each solid figure.

1. 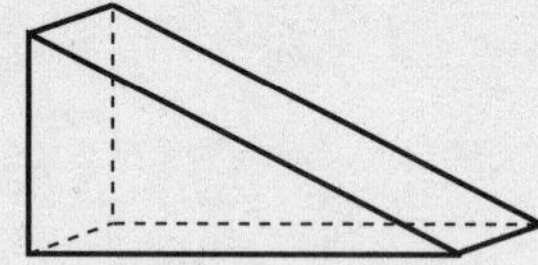

2. 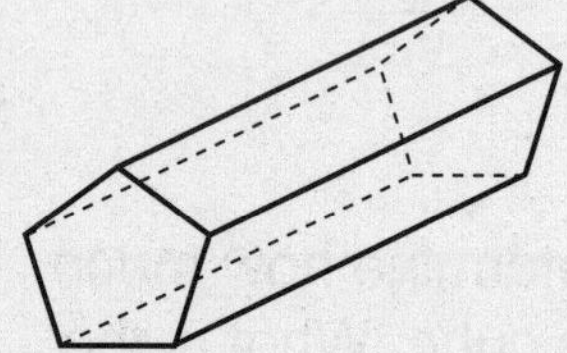

3. 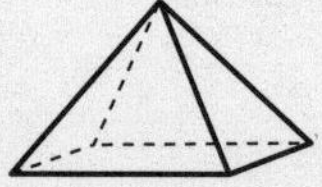

4. 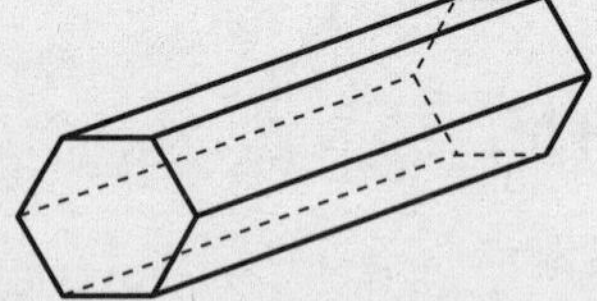

CCSS: 6.G.4

Directions: For questions 5 through 7, draw the net of each solid figure. Then find the surface area of the solid using its net.

5.

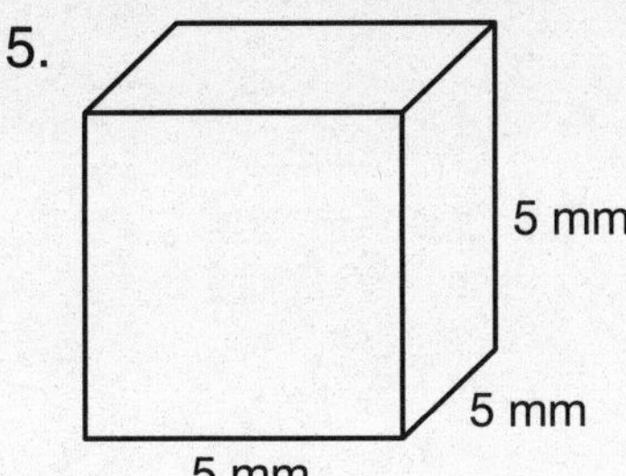

Surface area = ______________

6.

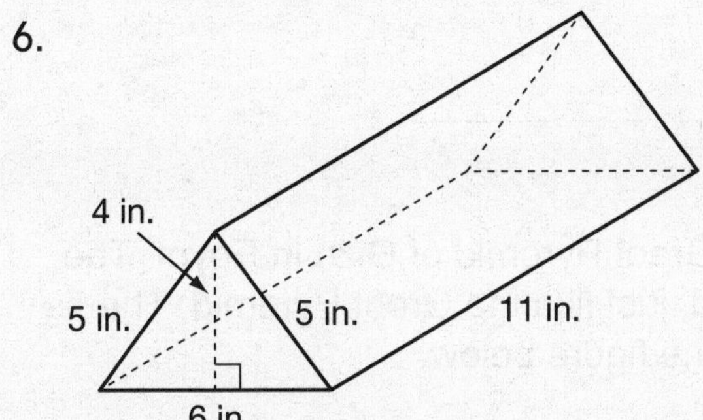

Surface area = ______________

7.

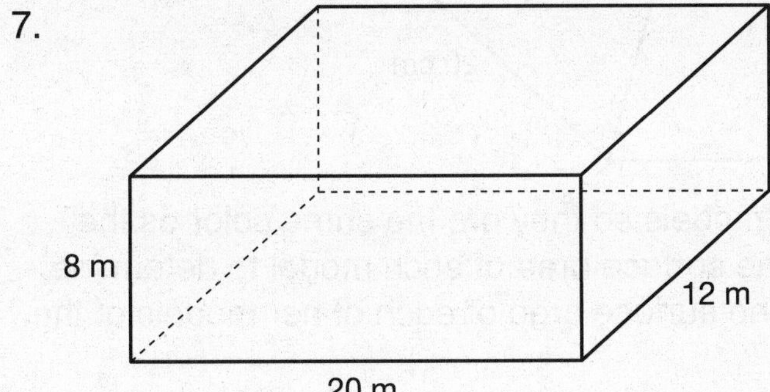

Surface area = ______________

8. A rice company has designed a box for a new product, as shown below.

What is the surface area of the box of rice? ____________

9. A gift shop owner sells scale models of the Great Pyramid of Giza in Egypt. The models are in the shape of a square pyramid, just like the Great Pyramid. The dimensions of the scale model are given in the figure below.

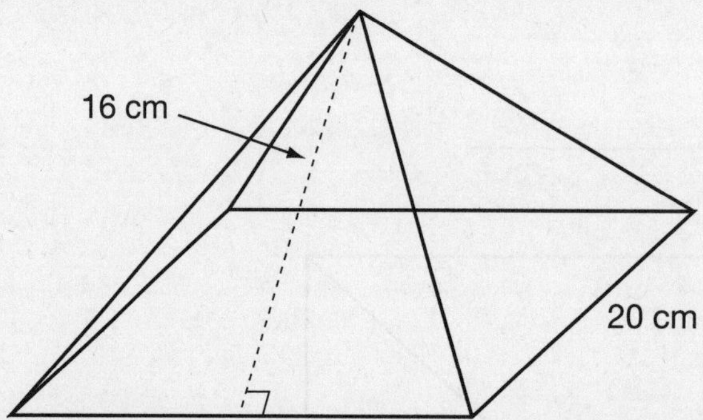

The owner of the gift shop paints the models so they are the same color as the actual pyramid. She needs to know the surface area of each model to determine how much paint she needs. What is the surface area of each of her models of the Great Pyramid?

____

Explain how you found your answer.

____

____

____

CCSS: 6.NS.8

Lesson 26: Coordinate Geometry

You can use a coordinate plane to determine the distance between two points with the same *x*- or *y*-coordinate.

Example

What is the distance between points *M* and *N* in the coordinate plane?

You can count the spaces between point *M* and point *N*. The distance between the points is 6.

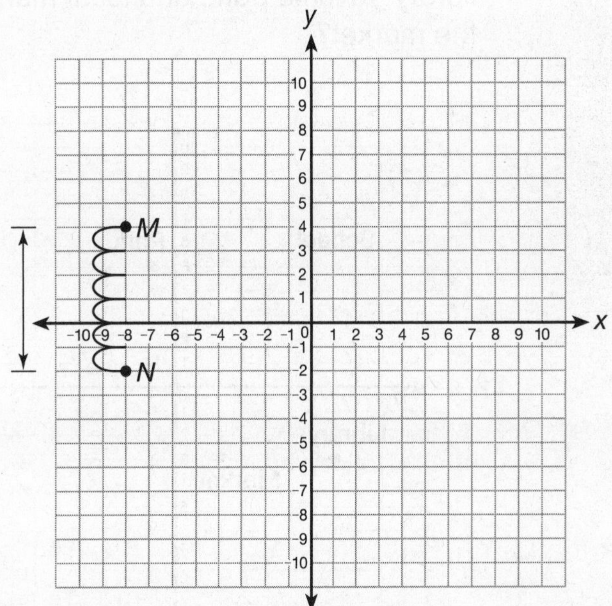

You can also use absolute value to find the distance between two points. The distance between two points is the absolute value of the difference between the coordinates that are different. The distance between the points (a, b) and (c, b) is $|a - c|$. The distance between the points (a, b) and (a, c) is $|b - c|$.

Example

What is the distance between $J(-7, 1)$ and $K(-4, 1)$?

Points *J* and *K* have the same *y*-coordinate. To find the distance between the points, find the absolute value of the difference of their *x*-coordinates.

$$|-7 - (-4)| = |-7 + 4| = |-3| = 3$$

The distance between *J* and *K* is 3.

Example

What is the distance between $R(-3, 4)$ and $S(8, 4)$?

Points *R* and *S* have the same *y*-coordinate. To find the distance between the points, find the absolute value of the difference of their *x*-coordinates.

$$|-3 - 8| = |-11| = 11$$

The distance between *R* and *S* is 11.

Maps have coordinates so that you can find locations easily.

Example

Willy makes a map to show the neighborhood around his house. Each unit on the plane represents a city block. He maps the locations of his house, school, library, favorite park, and local market. What is the distance from the library to the market?

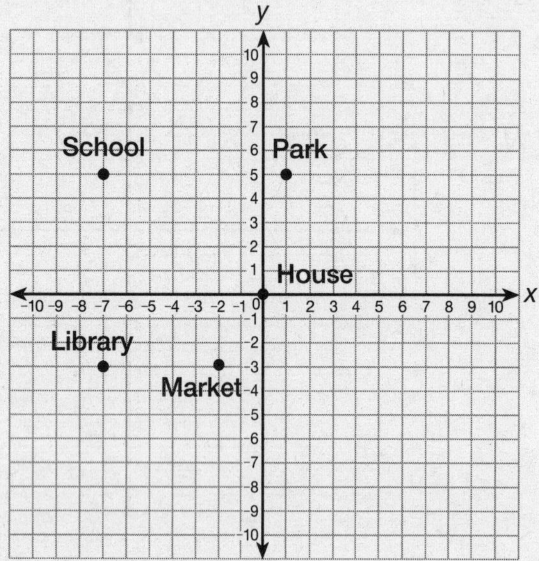

To find the distance between the points, you can count the number of units along the *x*-axis between them. You can also find the absolute value of the difference between the *x*-coordinates.

$$|-7 - (-2)| = |-7 + 2| = |-5| = 5$$

The distance from the library to the market is 5 city blocks.

Example

What is the distance from Willy's school to the library?

To find the distance between the points, you can count the number of units along the *y*-axis between them. You can also find the absolute value of the difference between the *y*-coordinates.

$$|5 - (-3)| = |5 + 3|$$
$$= |8|$$
$$= 8$$

The distance from Willy's school to the library is 8 city blocks.

CCSS: 6.NS.8

Practice

Directions: For questions 1 through 7, plot the points on the blank coordinate grid.

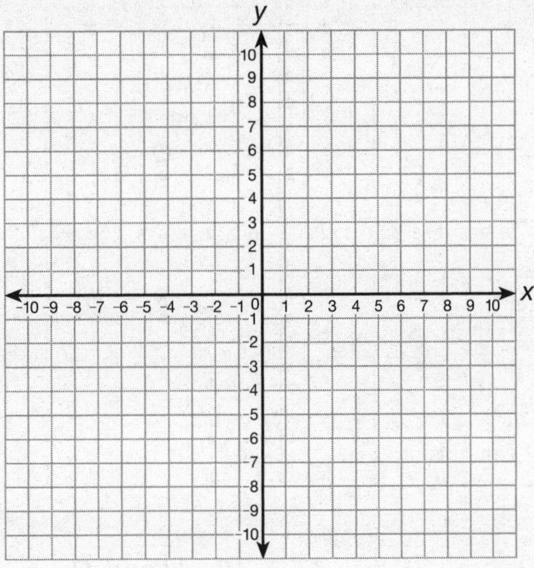

1. Plot point *A* at (−7, 7).

2. Plot point *B* at (−2, 4).

3. Plot point *C* at (2, 5).

4. Plot point *D* at (4, 0).

5. Plot point *E* at (−8, −6).

6. Plot point *F* at (0, −4).

7. Plot point *G* at (6, −3).

8. What is the ordered pair for the origin?

   A. (10, 0) C. (0, −4)

   B. (10, 10) D. (0, 0)

Directions: For questions 9 through 16, find the distance between the points.

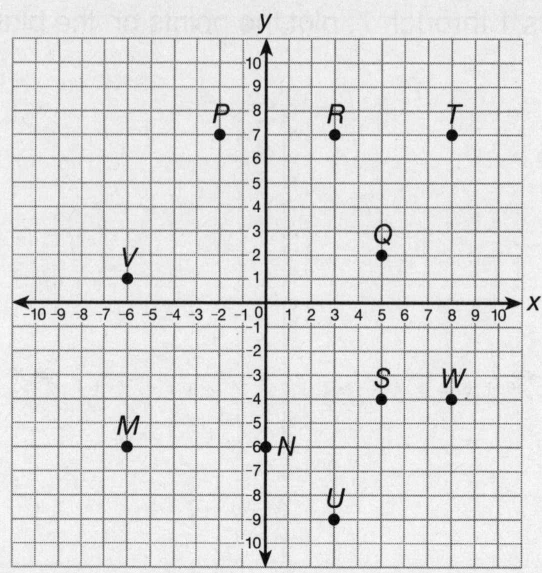

9. *P* and *T* ______________

10. *V* and *M* ______________

11. *S* and *W* ______________

12. *M* and *N* ______________

13. *U* and *R* ______________

14. *W* and *T* ______________

15. *R* and *T* ______________

16. *Q* and *S* ______________

17. What points, on the grid above, are 3 units away from the ordered pair (0, −9)?

vertically: ______________ horizontally: ______________

Directions: For questions 18 through 26, use the map.

Explorers of an undiscovered island create the following map of the island. They land their ship at the point at (6, 4) on the map.

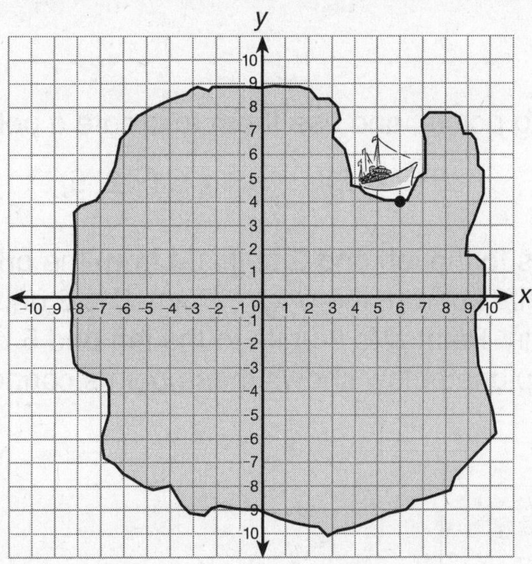

Directions: For questions 18 through 21, plot the locations as points on the grid.

18. Hill at (6, −2)

19. Pond at (−2, −2)

20. Apple tree at (−4, 4)

21. Sandy beach at (−4, −8)

Directions: For questions 22 through 25, find the distance between the locations.

22. Pond and hill ______________

23. Apple tree and ship ______________

24. Sandy beach and apple tree

____

25. Ship and hill ______________

26. Explain how you found the distance between the ship and the hill.

____

____

____

Lesson 27: Drawing Polygons

A **polygon** is a closed 2-dimensional figure made up of straight lines. You can create polygons on the coordinate grid.

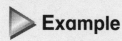

Example

Plot the following points, and use them to create a polygon.

$A(-8, 2)$ $B(-2, 2)$ $C(-2, 5)$ $D(-8, 5)$

Point *A* is 8 units to the left and 2 units up from the origin. Point *B* is 2 units to the left and 2 units up from the origin. Point *C* is 2 units to the left and 5 units up from the origin. Point *D* is 8 units to the left and 5 units up from the origin. The coordinate plane below shows these points connected with lines to form a rectangle.

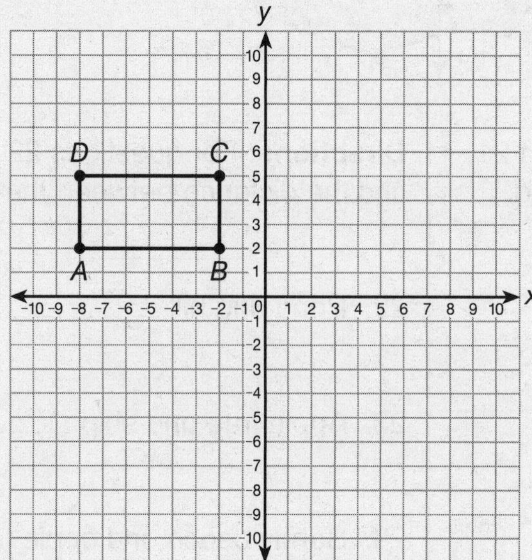

Example

Using the points in the coordinate plane from the previous example, find the distance from *A* to *B*, from *B* to *C*, from *C* to *D*, and from *D* to *A*.

Each pair of points has one identical coordinate in their ordered pairs. To find the distance between them, find the absolute value of the difference of the different coordinates.

A to *B*: $|-8 - (-2)| = |-6| = 6$
B to *C*: $|5 - 2| = |3| = 3$
C to *D*: $|-8 - (-2)| = |-6| = 6$
D to *A*: $|5 - 2| = |3| = 3$

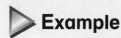

CCSS: 6.G.3

Example

A city bus follows a specific route. Each unit on the grid below represents one city block. The origin of the coordinate plane is the bus headquarters. The bus makes the following stops after leaving the bus station: college (4, 0), train station (4, 3), museum (−6, 3), stadium (−6, −3), and park (0, −3). Draw a polygon to show the bus route. Then determine the distances between the stops.

To create the polygon, plot the points and draw the route.

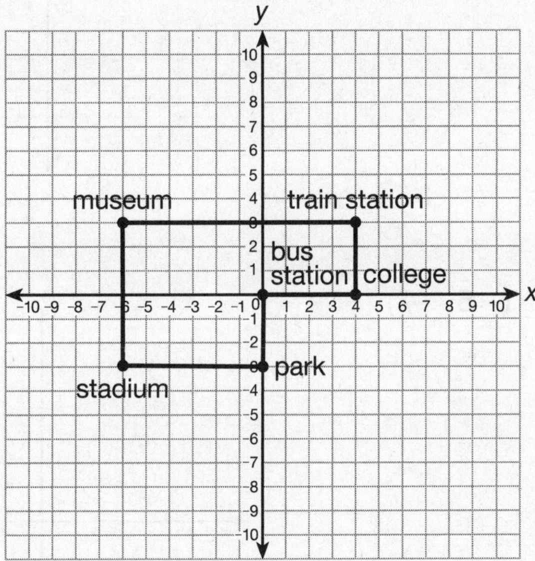

To determine the distances between the stops, find the absolute value of the difference between the different coordinates.

bus station to college: $|4 - 0| = |4| = 4$ blocks
college to train station: $|3 - 0| = |3| = 3$ blocks
train station to museum: $|4 - (-6)| = |10| = 10$ blocks
museum to stadium: $|3 - (-3)| = |6| = 6$ blocks
stadium to park: $|-6 - 0| = |-6| = |6| = 6$ blocks
park to bus station: $|3 - 0| = |3| = 3$ blocks

Practice

Directions: For questions 1 through 3, plot the given points to create a polygon.

1. *A* (−3, 7), *B* (7, 7), *C* (7, 1), *D* (−3, 1)

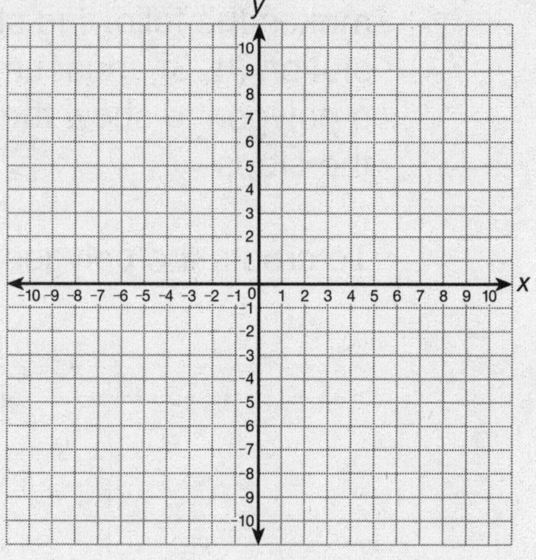

2. *F* (−5, 1), *G* (2, −1), *H* (2, −5), *J* (−5, −7)

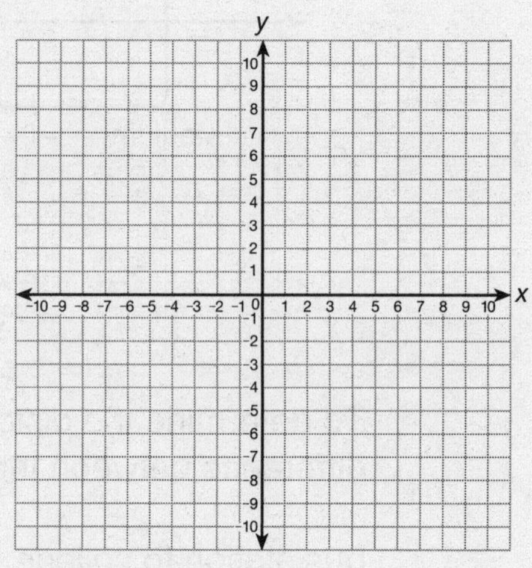

3. *L* (3, 7), *M* (5, 7), *N* (5, −4), *P* (−1, −4), *Q* (−1, −1), *R* (3, −1)

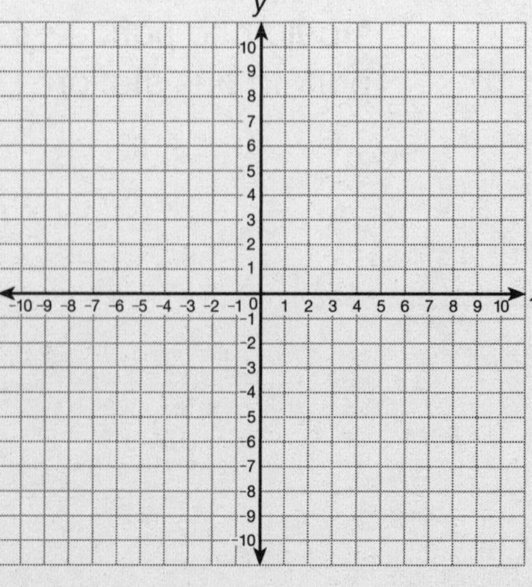

Directions: For questions 4 and 5, plot the given points to create a polygon. Then find the distances between the vertices.

4. An architect models an apartment building using a coordinate plane. Each unit on her model represents one apartment on a floor. She plots the front of the building with the following points:

   west floor: (1, 0); east floor: (8, 0); east roof: (8, 6); west roof: (1, 6)

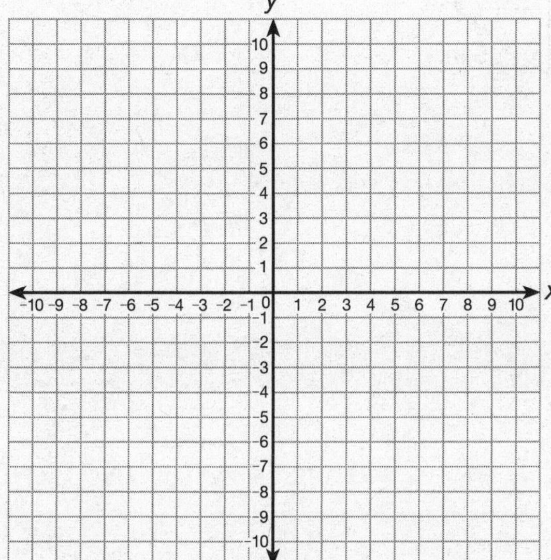

west floor to east floor: ______________

east floor to east roof: ______________

east roof to west roof: ______________

west roof to west floor: ______________

5. Melvin runs in his town's race. The start and finish of the race are at the same spot. Plot the route of the race using the following locations:

   start/finish: (2, 0); Park Place (2, 5); water station (−4, 5);

   bathrooms (−4, −3); pond (−2, −3); City Hall (−2, 0)

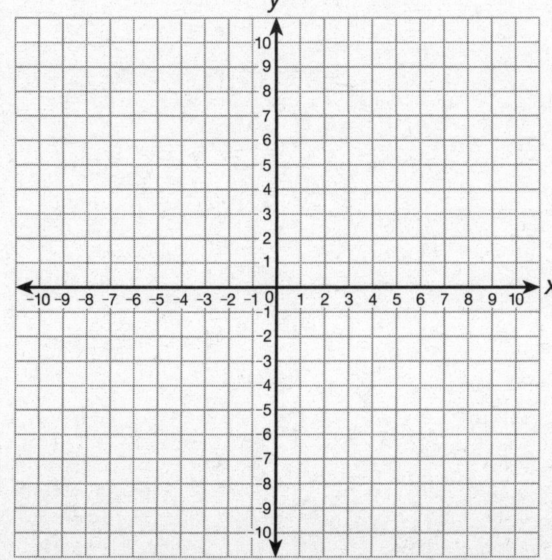

start/finish to Park Place: ______________

Park Place to water station: ______________

water station to bathrooms: ______________

bathrooms to pond: ______________

pond to City Hall: ______________

City Hall to start/finish: ______________

Unit 4 Practice Test

Read each question. Choose the correct answer.

1. Carrie buys the following box to send a gift to her nephew.

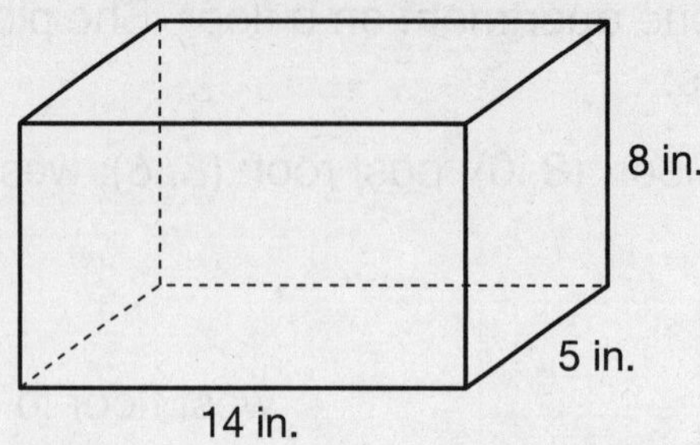

What is the volume of the box?

A. 112 in.^{3}

B. 444 in.^{3}

C. 480 in.^{3}

D. 560 in.^{3}

2. The following figure is a right triangle.

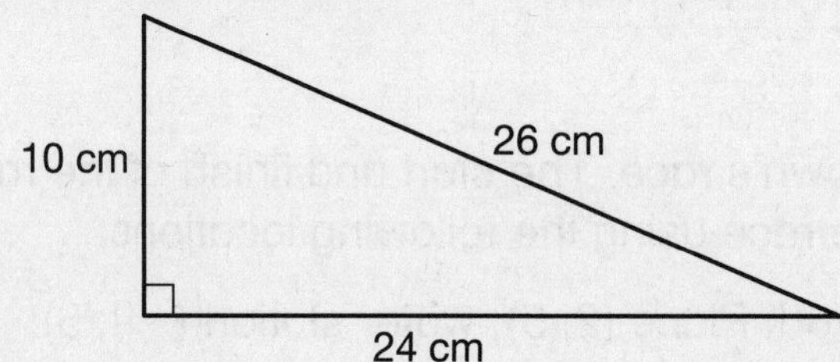

What is the area of the triangle?

A. 80 cm^2

B. 120 cm^2

C. 130 cm^2

D. 240 cm^2

3. Derek buys a doorstop with the following shape.

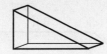

Which net can be used to construct the doorstop?

A.

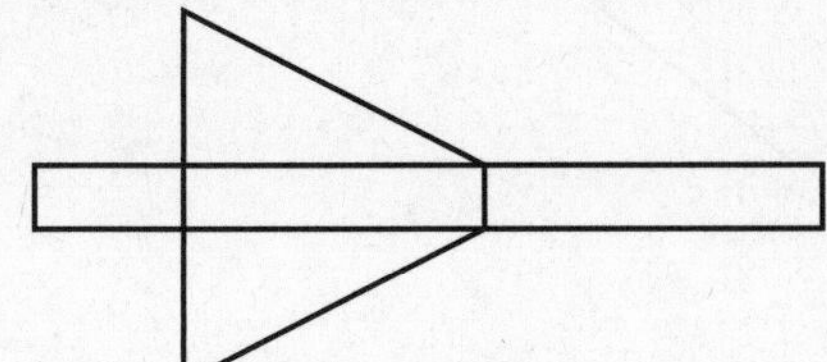

C.

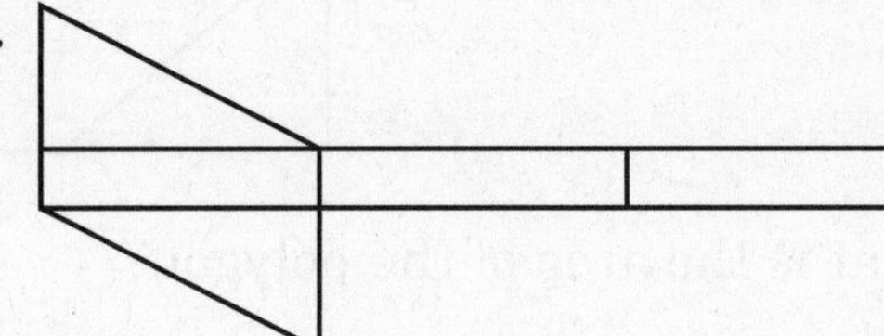

B.

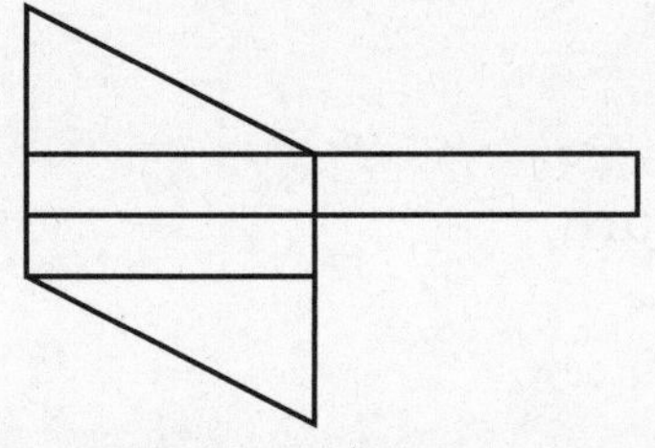

D.

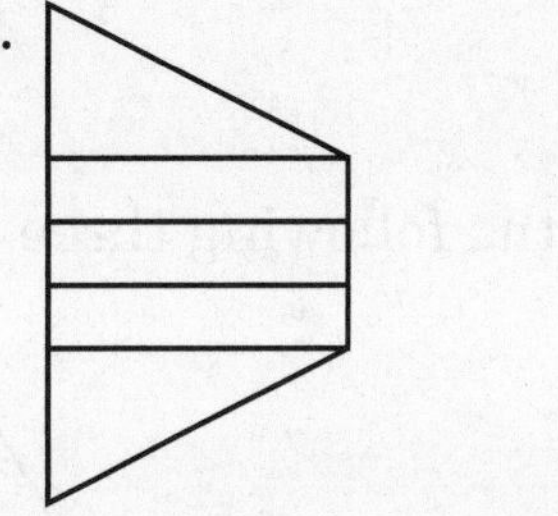

4. Plot the following points on the following coordinate grid.

$F(3, 2)$

$G(8, -1)$

$H(8, -5)$

$J(-2, -5)$

$K(-2, -1)$

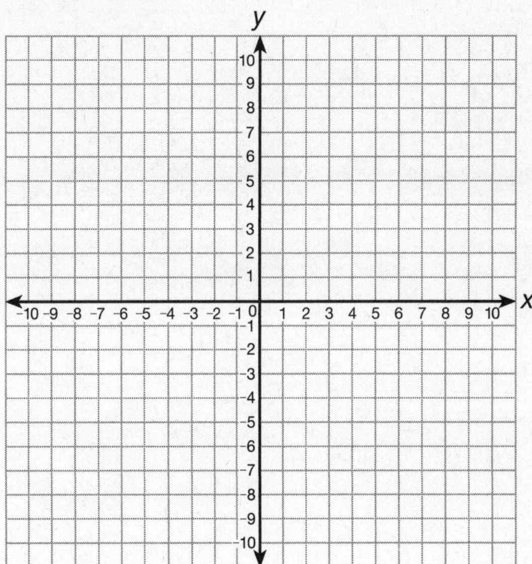

5. The following polygon is composed of a square and two triangles.

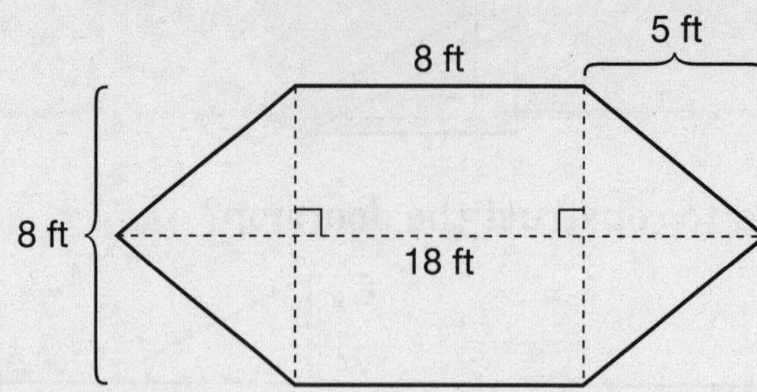

What is the area of the polygon?

____

6. Draw the net for the following three-dimensional figure.

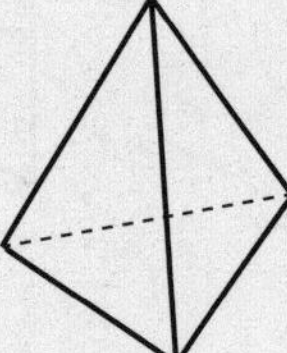

7. What is the volume of the following rectangular prism, in cubic units?

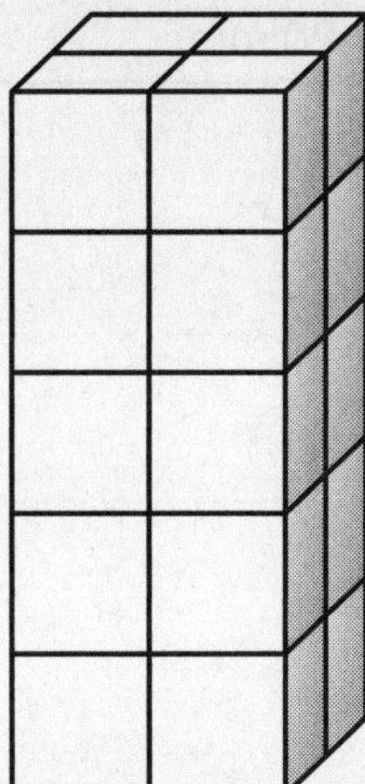

8. What is the area of the shaded part of the rectangle below?

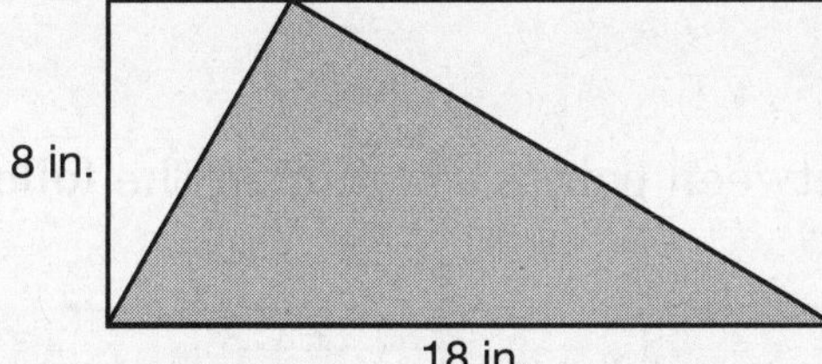

9. The following net represents a rectangular prism.

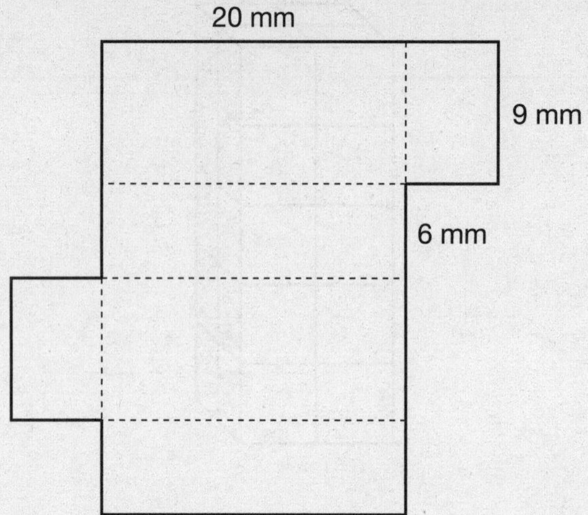

What is the surface area of the rectangular prism?

____

10. What is the distance between points *J* and *K* in the following coordinate plane?

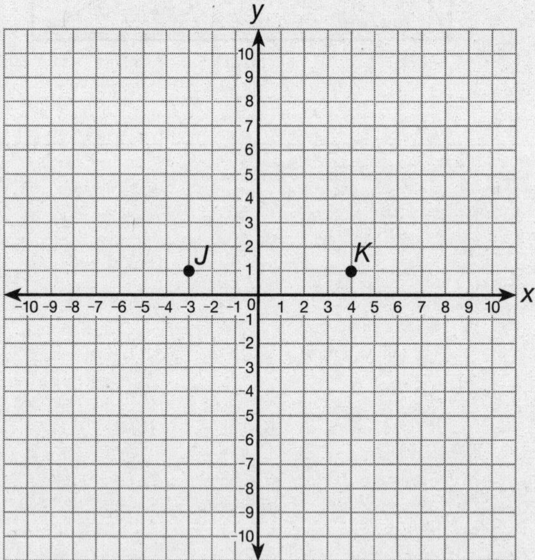

____

11. Yuri uses a tool called a T-square in his architecture class. The shape of the T-square is shown in the diagram below.

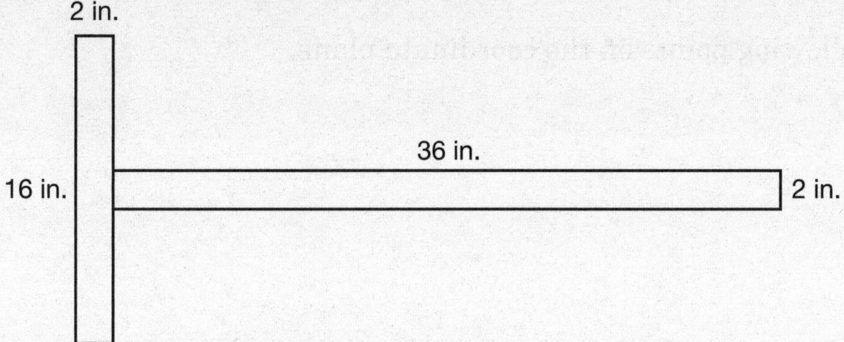

What is the area of Yuri's T-square? ______________________________

12. The following right rectangular prism is composed of cubes. Each cube has a length of $\frac{1}{2}$ in.

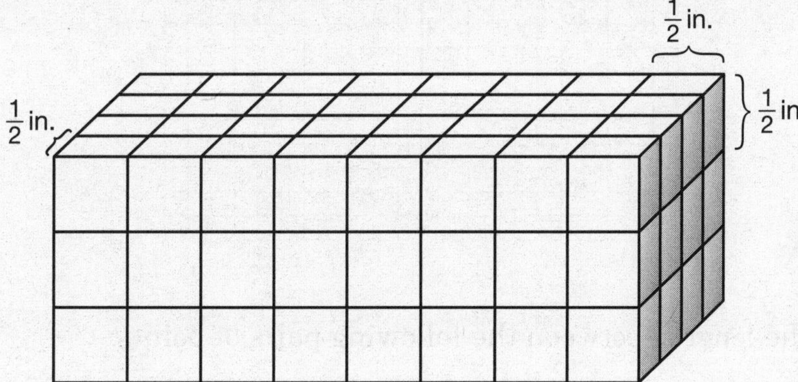

What is the volume of the rectangular prism? ______________________

13. Xavier holds a number cube with sides that are 1.5 cm long. He creates the following net for the number cube.

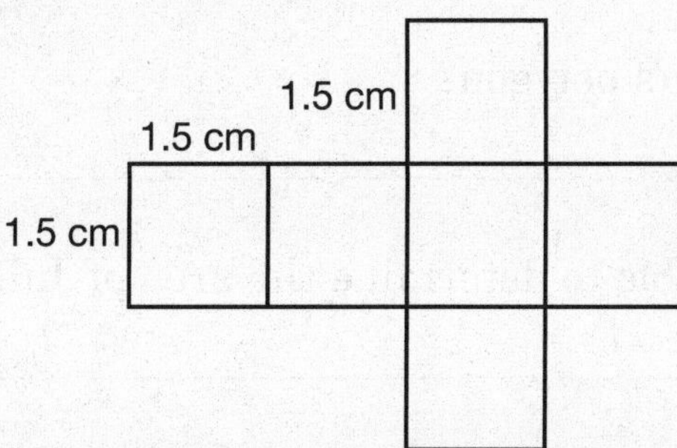

What is the surface area of Xavier's number cube? ____________________

14. Julio describes the shape of a polygon by writing down the ordered pairs for the vertices.

Part A
Plot the following points on the coordinate plane.

A (−4, −8)

B (5, −8)

C (5, 5)

D (−4, 5)

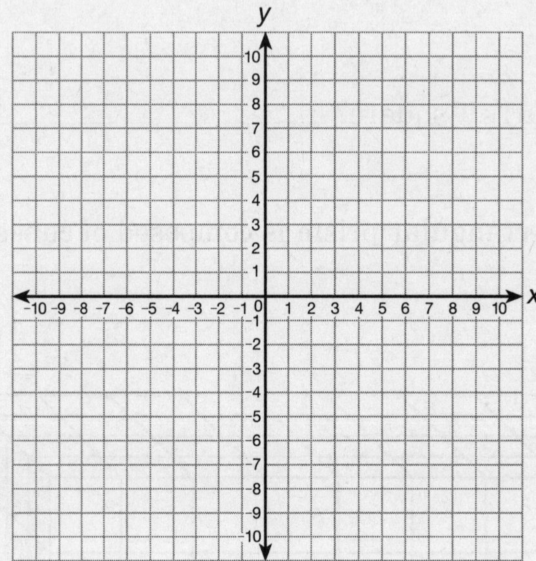

Part B
Determine the lengths between the following pairs of points.

A and *B*: ______________

B and *C*: ______________

C and *D*: ______________

A and *D*: ______________

Part C
What is the area of Julio's polygon?

____

Explain how you were able to determine the area of Julio's polygon.

____

____

____

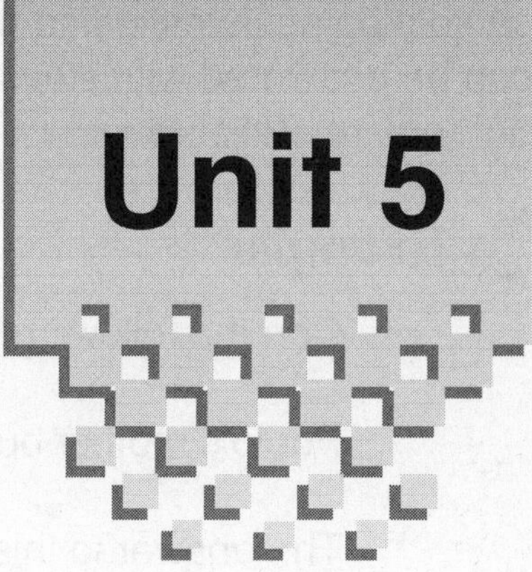

Statistics and Probability

Data analysis is used in more ways than you might think. Farmers may consider data about rainfall, temperature, and yield when planting crops. City planners look at traffic data when they think about building new roads. There are many ways to represent and describe the data in a set as well. The different representations show different aspects of the data, such as the most frequent values or the median.

In this unit, you will learn to determine whether a question can be answered with statistics. You will use different measures to describe the center of a data set, including the mean, median, mode, range, interquartile range, and mean absolute deviation. You will plot data sets using a number line, dot plot, box plot, and histogram. Finally, you will learn about the different ways to describe a data set.

In This Unit

Statistical Questions

Measuring Data Sets

Plotting Data Sets

Describing Data Sets

Lesson 28: Statistical Questions

Statistics can be used to answer many types of questions. However, some questions do not need statistics to be answered. Statistics are needed to solve a question if the subject being investigated has variability. If there is variability in the data, the question can be answered with statistics. If the question has a simple numerical answer, it may not require statistics.

Example

Can the following question be answered with or without statistics?

Paulo's mother buys a new car. How much did she spend on her new car?

The answer to this question will be an exact number. It will represent the measurement of the cost of the car. There is no variability in the cost. Therefore, statistics are not required to answer the question.

Example

Can the following question be answered with or without statistics?

A car dealer has 45 cars for sale. How much does the dealer charge for the cars?

The answer to this question will likely be a range of values. There is not one numerical answer. You would expect variability in the data. Therefore, the question can be answered with statistics.

Example

Will the following question be answered with or without statistics?

Nathan is the tallest student in his class. How tall is Nathan?

There is no variability in the data: Nathan will have only one height. Therefore, the question can be answered without statistics.

Practice

Directions: For questions 1 through 8, determine whether statistics are needed for the answer. Write *yes* if it the question can be answered with statistics. Write *no* if it can be answered without statistics.

1. The nature park has 18 monkeys. What is the weight of the monkeys?

   ____

2. Jupiter is the largest planet in the solar system. How many moons does Jupiter have?

   ____

3. Nadine got the highest score on the quiz. What score did Nadine get?

   ____

4. A meteorologist measures the weather every day. How many rainy days are there each month in a year?

   ____

5. Eugene is shopping for a new computer. How much do computers cost at the store?

   ____

6. A bank charges a flat fee for a withdrawal from its ATM. How much does the bank charge?

   ____

7. Geoffrey measures his foot at the shoe store. What size shoe does he wear?

   ____

8. An elementary school has 10 classes. How many students are in each class?

   ____

Lesson 29: Measuring Data Sets

To measure a data set, you can use the mean, median, and mode. Each of these measures summarizes a data set with a single number.

Mean

The **mean** is the sum of the numbers in a data set divided by how many numbers there are in the set. The mean is often called the **"average."**

$$\text{mean} = \frac{\text{sum of the numbers}}{\text{how many numbers}}$$

Example

A train has five cars with 45, 51, 48, 39, and 42 passengers in them. What is the mean number of passengers per train car?

First, add the five numbers.

$$45 + 51 + 48 + 39 + 42 = 225$$

Now substitute the numbers into the formula, and find the mean.

$$\text{mean} = \frac{\text{sum of the numbers}}{\text{how many numbers}}$$

$$= \frac{225}{5}$$

$$= 45$$

The mean number of passengers per train car is 45.

Median

The **median** is the **middle** number in a data set arranged from least to greatest (or greatest to least).

 **Example**

The following list shows the number of customers who ordered macaroni at Dina's Diner each day last week.

11, 15, 8, 18, 10, 15, 12

What is the median number of customers who ordered macaroni?

Arrange the numbers in order from least to greatest. Find the middle number.

8, 10, 11, (12), 15, 15, 18

The median number of customers who ordered macaroni is 12.

Mode

The **mode** is the number that appears **most often** in a data set. There can be one or more modes in a data set. If all the numbers appear the same number of times, there will be no mode at all.

 **Example**

Stephen played ten games of baseball and had the following numbers of at-bats.

4, 5, 3, 6, 3, 4, 4, 1, 5, 4

What is the mode number of at-bats?

Find the number that appears most often.

(4), 5, 3, 6, 3, (4), (4), 1, 5, (4)

The mode number of at-bats is 4.

 **TIP:** To find the median of a data set with an even number of numbers in a data set, find the two middle numbers. Add the two middle numbers, then divide by 2 to find the median.

To measure a data set, you can also use the range, interquartile range, and the mean absolute deviation. Each of these measures describes how a data set varies with a single number.

Range

The **range** is the difference between the least number and the greatest number in a data set.

 **Example**

Cody's Crab Shack had the following number of customers last week. What is the range of the number of customers in Cody's Crab Shack?

47, 38, 55, 68, 40, 33, 55

Subtract the smallest number in the data set, 33, from the greatest number, 68.

$68 - 33 = 35$

The range number of customers in Cody's Crab Shack is 35.

Interquartile Range

The **interquartile range** is the difference between the first and third **quartiles** of a data set. To find the **first quartile**, find the median of the lower half of a data set. To find the **third quartile**, find the median of the upper half of a data set.

 **Example**

What is the interquartile range of the following data set?

5, 15, 25, 9, 25, 12, 25, 30, 20, 21

First, put the numbers in numerical order.

5, 9, 12, 15, 20, 21, 25, 25, 25, 30

The median of the data set is $\frac{20 + 21}{2} = 20.5$.

The lower half of the data set is 5, 9, 12, 15, and 20. The median of this lower half is 12. The first quartile is 12.

The upper half of the data set is 21, 25, 25, 25, and 30. The median of this upper half is 25. The third quartile is 25.

The interquartile range is the difference between the first and third quartiles: $25 - 12 = 13$. The interquartile range of the data set is 13.

Mean Absolute Deviation

The **mean absolute deviation** is the average amount by which the measurements in a data set vary from the mean. To find the mean absolute deviation, first find the absolute value of the difference between each data point and the mean. Then find the average of those deviation amounts.

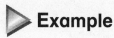

Example

Charlene gets the following scores during the first five holes of a golf tournament. Find the mean absolute deviation of her scores.

4, 5, 2, 8, 6

First, calculate the mean of the data set.

$$\text{mean of the data set} = \frac{\text{sum of the numbers}}{\text{how many numbers}}$$

$$= \frac{4 + 5 + 2 + 8 + 6}{5}$$

$$= \frac{25}{5}$$

$$= 5$$

The mean score of the data set is 5. Now find the absolute value of the difference between each data point and the mean.

hole 1: $|5 - 4| = 1$
hole 2: $|5 - 5| = 0$
hole 3: $|5 - 2| = 3$
hole 4: $|5 - 8| = 3$
hole 5: $|5 - 6| = 1$

Finally, find the average of these numbers.

$$\text{mean of the deviation points} = \frac{\text{sum of the numbers}}{\text{how many numbers}}$$

$$= \frac{1 + 0 + 3 + 3 + 1}{5}$$

$$= \frac{8}{5}$$

$$= 1.6$$

The mean absolute deviation of the data set is 1.6.

TIP: The greater the mean absolute deviation, the less the mean accurately represents the data set. A small mean absolute deviation means that the data points are close to the mean.

Practice

Directions: For questions 1 and 2, find the mean, median, and mode of the data set. If there is no mode, write *none*.

1. A fishing boat caught the following weights of fish, in pounds, on 7 fishing trips at Lake Michigan.

   90, 53, 50, 59, 63, 50, 97

   mean: ____________

   median: ____________

   mode: ____________

2. As of 2010, the numbers of U.S. representatives for the six New England states are given below.

   Connecticut: 5, Massachusetts: 10, Maine: 2,
   New Hampshire: 2, Rhode Island: 2, Vermont: 1

   mean: ____________

   median:____________

   mode: ____________

3. Eduardo's test scores are 85, 93, 85, 94, and 88. What is the median of his test scores?

   A. 85

   B. 86.5

   C. 88

   D. 89

4. Sloan read seven books over her summer vacation. The numbers of pages in the books are shown in the following list.

   212, 340, 288, 255, 340, 313, 289

   What is the mean number of pages of her books?

   A. 288 C. 291

   B. 289 D. 340

Directions: For questions 5 and 6, find the range and interquartile range of the data set.

5. As of 2010, the 10 tallest buildings in the world had the following numbers of stories:

   160, 101, 101, 88, 88, 66, 110, 103, 88, 88

   range: ____________

   interquartile range: ____________

6. A scientist counts the numbers of eggs laid by 8 different leatherback turtles on the beach. The numbers of eggs from the turtles are listed below.

   88, 92, 80, 80, 82, 93, 96, 95

   range: ____________

   interquartile range: ____________

7. A store owner wants to determine the average age of the customers in her store. The following list shows the ages of the 6 customers in her store.

   16, 49, 19, 15, 15, 18

   What is the mean absolute deviation?

   ____

   Explain how you found the mean absolute deviation.

   ____

   ____

   ____

   ____

Lesson 30: Plotting Data Sets

Data can be plotted in a variety of different ways. Depending on the way that the data are plotted, a graph can show the way that the points are distributed.

Plots on a Number Line

You can use a number line to plot the points in a data set.

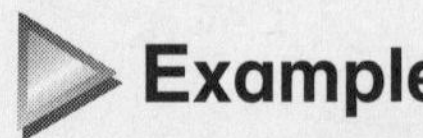

Example

Dana has three sisters. Her sisters are 2, 5, and 14 years old. She also has a brother who is 9 years old. Plot the ages of Dana's brother and sisters on the following number line.

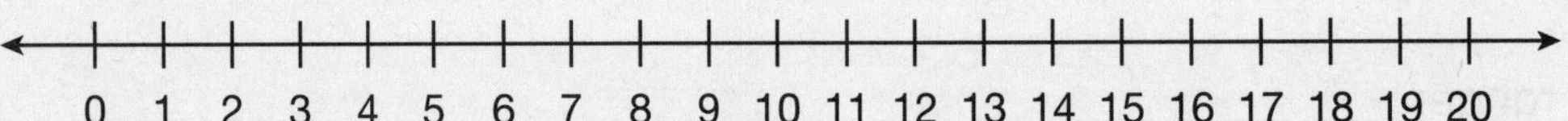

The data set includes the ages of Dana's sisters and brother. The siblings are 2, 5, 9, and 14 years old. To plot the points from that set on a number line, draw a point by each of those numbers.

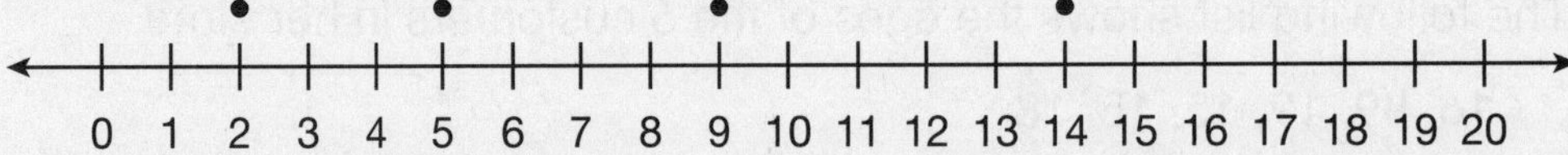

The number line represents the ages of Dana's brother and sisters.

CCSS: 6.SP.2, 6.SP.4

Dot Plots

A **dot plot** shows the number of times each value in a data set occurs. A number line and dots are used to organize the data. A dot plot is useful when there are several identical points in the same data set.

Example

The list shows the high temperatures (in °F) in New York City in August 2010. Plot the temperatures in a dot plot.

79, 83, 87, 93, 91, 86, 84, 86, 91, 90, 91, 77, 81, 79, 79, 87,

88, 84, 88, 89, 82, 81, 71, 71, 70, 82, 78, 83, 93, 91, 95

To add the points in a dot plot, draw a dot by each number. If there is more than one point with the same value, draw one dot above the other.

August 2010 High Temperatures in New York City

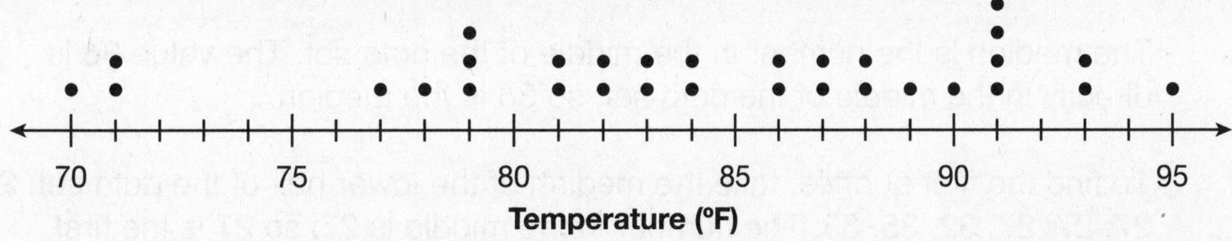

The dot plot shows the distribution of points from the lowest number, 70, to the highest number, 95. Because the most dots are above 91, you can see that there was a high temperature of 91 degrees the greatest number of days.

Box Plots

A **box plot** shows the range of values in a data set, including the minimum and maximum value. It also represents how the values are distributed by showing the median, first quartile, and third quartile of the data set.

Example

A city bus company counts the number of passengers who ride on 15 different buses one morning.

32, 25, 40, 45, 27, 42, 35, 48, 27, 36, 27, 40, 39, 35, 41

First, put the numbers in order from least to greatest.

25, 27, 27, 27, 32, 35, 35, 36, 39, 40, 40, 41, 42, 45, 48

The minimum value is 25, and the maximum value is 48. These will be the end points of the box plot.

The median is the number in the middle of the data set. The value 36 is directly in the middle of the data set, so 36 is the median.

To find the first quartile, take the median of the lower half of the data set: 25, 27, 27, 27, 32, 35, 35. The number in the middle is 27, so 27 is the first quartile.

To find the third quartile, take the median of the upper half of the data set: 39, 40, 40, 41, 42, 45, 48. The number in the middle is 41, so 41 is the third quartile.

Use the first and third quartiles to form the borders of the box. Draw a line in the box to show the median.

Number of Passengers in City Buses

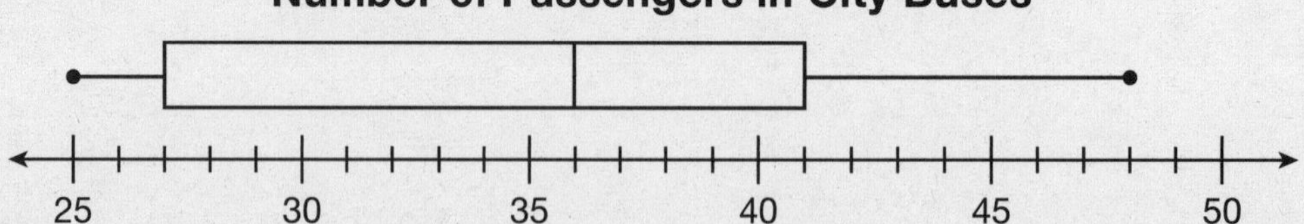

CCSS: 6.SP.2, 6.SP.4

Histograms

A **histogram** shows continuous data in a set. The horizontal axis is labeled using intervals, and the bars are always connected to each other. The height of each bar represents the frequency.

Example

Tyler recorded the distances traveled by the first 25 participants in a 100-mile bike tour after 5 hours. Plot the following data points in a histogram.

54, 69, 72, 44, 65, 18, 88, 91, 75, 62, 48, 51, 83, 70, 65, 54, 76, 63, 34, 82, 67, 77, 65, 59, 79

First, put the numbers in order from least to greatest.

18, 34, 44, 48, 51, 54, 54, 59, 62, 63, 65, 65, 65, 67, 69, 70, 72, 75, 76, 77, 79, 82, 83, 88, 91

To plot these data points in a histogram, you can choose intervals. Tyler's data points begin at 18 and go to 91. To show the data, let the intervals of the histogram be in 10s of miles; 1–10, 11–20, and so on until 91–100.

The height of each bar should represent the number of values in the interval. For example, if there is one value in an interval, draw a bar that goes to 1 on the y-axis.

Tyler's results are shown in the following histogram.

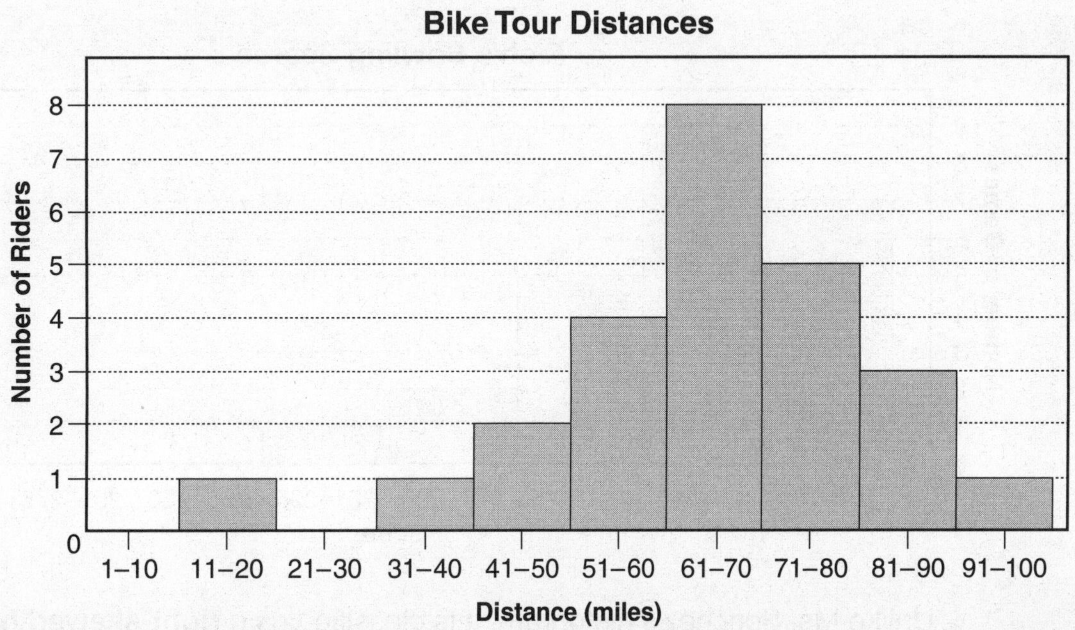

Example

Ms. Sanchez graded the tests for her 28 students and made the following histogram to show the distribution of the scores. Describe the distribution of the data in her histogram.

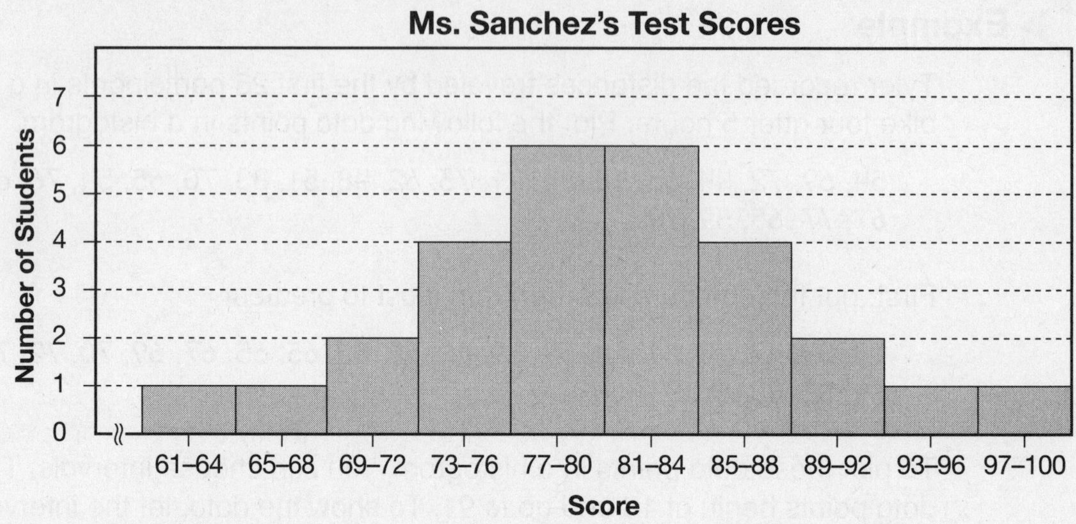

Ms. Sanchez's histogram has left and right sides mirror images of each other. It is a **bell-shaped histogram** because it is in the shape of a bell.

Example

Rich made the following histogram to show his bowling scores from his last 25 games. Describe the distribution of the data in his histogram.

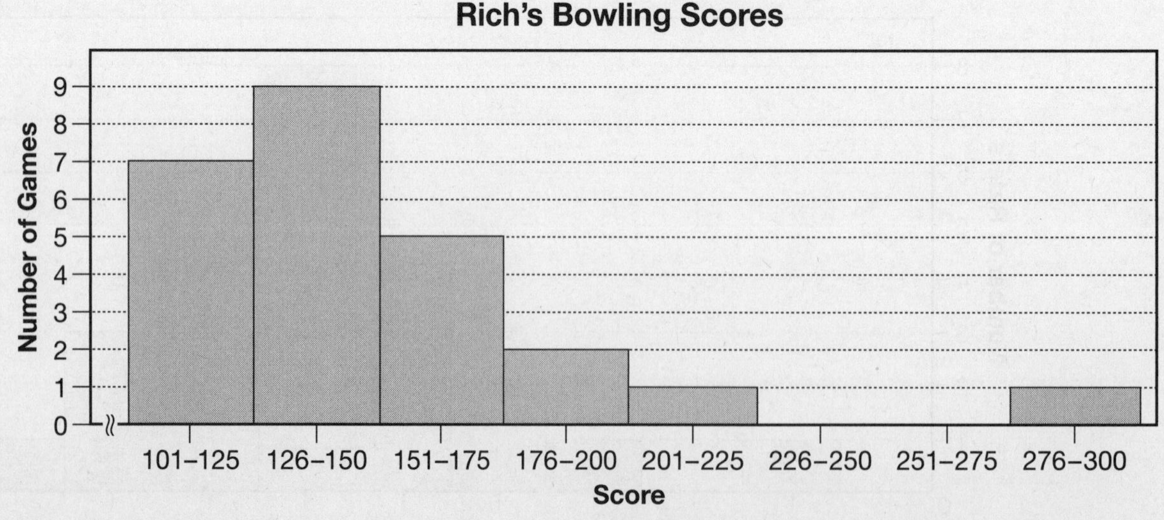

Unlike Ms. Sanchez's histogram, it is classified as a **right-skewed histogram** because the right side of the histogram is longer.

CCSS: 6.SP.2, 6.SP.4

Practice

Directions: Use the following information for questions 1 through 3.

To prepare for her race, Rachael ran the following number of miles each day last week.

Monday: 8, Tuesday: 10, Wednesday: 6, Thursday: 12, Friday: 14

1. Plot the distances that Rachael ran on the following number line.

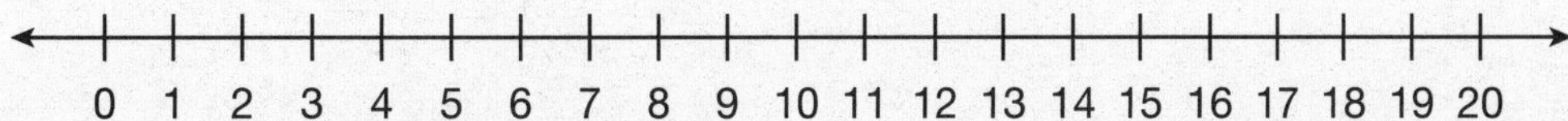

2. What is the greatest number of miles that Rachael ran in one day? ____________

3. What is the range of miles that Rachael ran during the week? ____________

Directions: Use the following information for questions 4 through 6.

Mr. Mantle asked the 20 students in his class how many hours they spent studying last week. The results are given below.

6, 9, 3, 16, 4, 5, 9, 15, 5, 6, 8, 9, 15, 9, 12, 14, 15, 10, 19, 12

4. Plot Mr. Mantle's data set in the following dot plot.

Hours Spent Studying

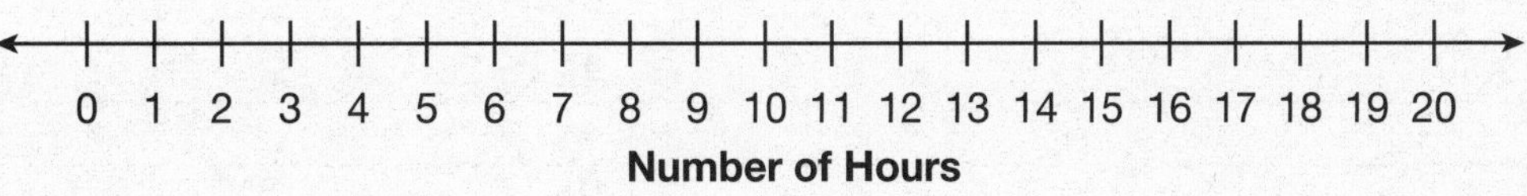

5. How many students in Mr. Mantle's class spent 15 hours studying last week?

____

6. What is the most common number of hours spent studying last week by Mr. Mantle's students?

____

Directions: Use the following information for questions 7 through 9.

Christopher counted the number of hours he worked each week for the first 15 weeks of the year.

20, 28, 30, 30, 30, 32, 35, 36, 38, 40, 40, 41, 42, 42, 44

7. Plot Christopher's data set in the following box plot.

Hours Christopher Worked Each Week

Number of Hours Worked

8. What is the median number of hours that Christopher worked in a week? ________

Directions: Use the following information for questions 9 through 11.

Ms. Ng had 18 customers in her garage this week. The amount she charged each customer depended on the work done to the customer's car. The amounts are shown below, in dollars.

245, 450, 380, 290, 895, 550, 210, 275, 320, 585, 255, 420, 180, 620, 305, 420, 195, 230

9. Plot Ms. Ng's data set in the following histogram.

Amounts Charged by Ms. Ng

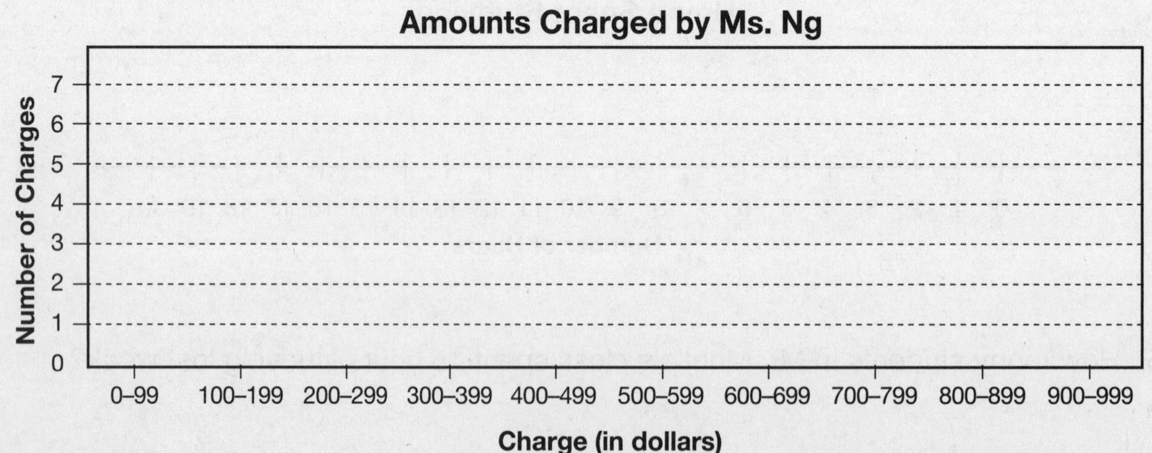

10. How many of Ms. Ng's customers were charged between $200 and $299? ________

11. How many of Ms. Ng's customers were charged more than $700? ________

CCSS: 6.SP.2, 6.SP.4

Directions: Use the following information for questions 12 through 14.

There were 32 births in the county hospital last month. The weights of the babies, in pounds, are given in the following table.

Weights of Babies in County Hospital (in pounds)							
8.4	4.7	7.7	8.0	7.5	5.4	8.8	7.5
6.9	7.2	3.9	11.1	5.5	10.1	6.7	9.8
7.8	6.2	9.3	8.3	6.7	7.5	10.3	4.8
8.1	7.9	6.4	9.6	5.9	6.6	8.8	7.3

12. Plot the hospital's data set in the following histogram.

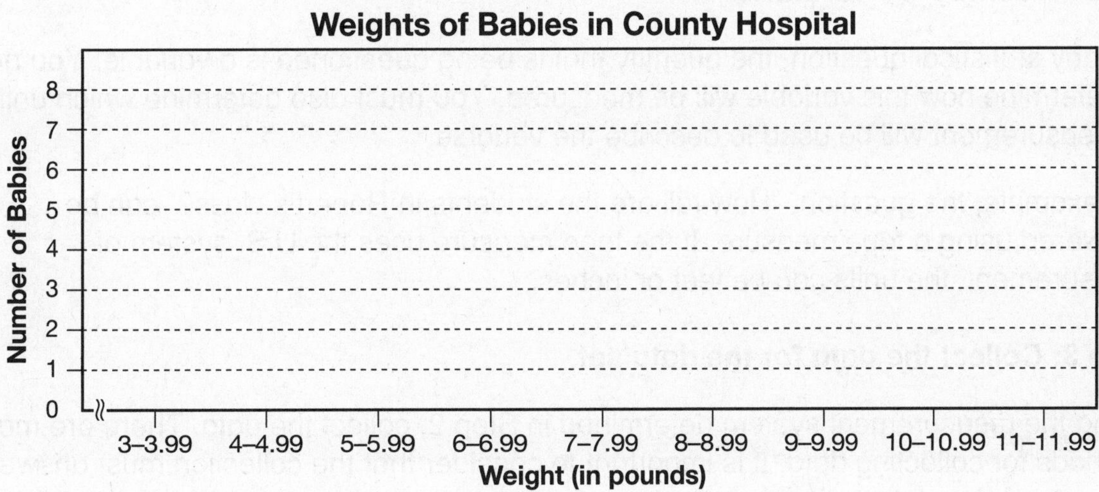

13. According to the hospital's data, between which two whole numbers is the most common weight for a baby?

14. Classify the histogram that you created. ____________

    Describe how you used the distribution of the data to classify the histogram.

    ____

    ____

    ____

Lesson 31: Describing Data Sets

To solve a statistical question, you need to follow the following steps.

Step 1: Identify whether the question requires statistics.

As covered in Lesson 28, questions may or may not require statistics to be solved. If the answer to the question will not include variability in the data, then statistics may not be required.

For example, "How tall is Robert?" is not a statistical question because there is no variability. The question, "How tall are the students in Robert's class?" *is* a statistical question because there will be variability in the heights of the students.

Step 2: Identify the variable.

For any statistical question, the quantity that is being questioned is a variable. You need to determine how this variable will be measured. You must also determine which units of measurement will be used to describe the variable.

For example, the question, "How tall are the students in Robert's class?" can be answered using a tape measure. If the tape measure uses the U.S. system of measurement, the units can be feet or inches.

Step 3: Collect the data for the data set.

Using the measurement system determined in Step 2, collect the data. There are many methods for collecting data. It is important to consider that the collection must answer the statistical question directly.

Consider the following question: "What is the average amount of rain that falls in the park each day?" The collection method must be accurate and consistent to answer the question. For example, a container must be located in a place that measures the rainfall accurately. Placing the container below a tree could influence the results.

A data set will be more accurate if more values are collected for the set. For example, you could collect rainfall every day for a week and answer the question. However, it would be more accurate to measure the rainfall every day for a year. The larger data set would provide a better solution to the statistical question.

Step 4: Calculate the measures of the data set.

There are several measures that can be used to summarize a data set. Those measures include the mean, median, mode, range, interquartile range, and mean absolute deviation. Each of these measures represents a different way to interpret the data.

The mean, median, and mode show the **measures of center** of the data. Measures of center show a way that the average of the data can be represented. The range, interquartile range, and mean absolute deviation are **measures of variability**. Measures of variability show how the data varies within the set. Measures of center and variability both represent the data set in different ways.

Step 5: Display the data.

There are many ways to display data. For example, you can use a dot plot, a box plot, or a histogram. Each type of display has its own benefits. For example, a dot plot can easily show the most common values in a data set. A box plot can show the range, median, and quartiles of a data set. A histogram shows continuous data using intervals. You can also represent the same data set using multiple displays.

Step 6: Analyze the data.

Once the data from the data set are displayed, you can analyze the data. Look for patterns or deviations in the data. For example, if the display is a histogram, what is the shape of the histogram? Is the histogram left- or right-skewed? Does the data show a bell-shaped distribution of values? Is there anything unusual that stands out in the data?

Consider how the data were collected. Determine whether the context of the data collection affected the results.

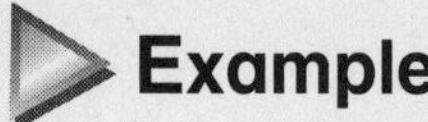

Example

Follow the steps to answer the following question.

What are the ages of the customers at a G-rated movie?

Step 1: Identify whether the question requires statistics.

It is unlikely that everyone in the movie theater is the same age. Therefore, this question will likely include variability in the data. It is a statistical question.

Step 2: Identify the variable.

The information that you will need to find is the ages of the customers at the movie. The unit of measurement to use for this data is a year. Knowing the exact age in terms of days, weeks, or months is probably not necessary.

The simplest way to measure the data is to simply ask the customers for their ages. No additional tools are needed for measurement.

Step 3: Collect the data for the data set.

The data can be collected from different places. For example, the person collecting the data could ask the customers for their ages outside the ticket window, by the entrance, at the snack bar, or inside the theater. A survey conducted outside the ticket window may ask people on the street who will not be customers of the G-rated movie.

A person who is trying to collect the data may not be able to ask every customer for his or her age. The more customers who participate in the survey, the more accurate the results will be. The following table shows the ages of the 32 customers entering the G-rated movie.

Ages of Customers at Movie (in years)							
5	8	28	30	6	11	33	6
52	3	39	8	34	12	7	42
33	45	11	4	6	44	68	12
8	6	40	27	35	10	5	37

Step 4: Calculate the measures of the data set.

Before any of the measures can be calculated, put the ages in order from least to greatest.

3, 4, 5, 5, 6, 6, 6, 6, 7, 8, 8, 8, 10, 11, 11, 12, 12, 27, 28, 30, 33, 33, 34, 35, 37, 39, 40, 42, 44, 45, 52, 68

The sum of the values is 715. There are 32 values. The mean age is $\frac{715}{32}$, or about 22 years old. There is an even number of values in the set. The two middle numbers are both 12. Because $\frac{12 + 12}{2} = 12$, the median age is 12. The age 6 appears more often than any other value in the data set, so the mode is 6 years old. The range is 68 − 3, or 65.

The first quartile is the median of the lower half of values, 6.5. The third quartile is the median of the upper half of values, 36. Because 36 − 6.5 = 29.5, the interquartile range is 29.5.

To find the mean absolute deviation, find the absolute value of the difference between each value and 22. The differences are 19, 18, 17, 17, 16, 16, 16, 16, 15, 14, 14, 14, 12, 11, 11, **10, 10**, 5, 6, 8, 11, 11, 12, 13, 15, 17, 18, 20, 22, 23, 30, and 46. The mean absolute derivation is the mean (average) of those numbers, which is equal to about 15 years.

Step 5: Display the data from the data set.

A histogram can display the different age groups using intervals.

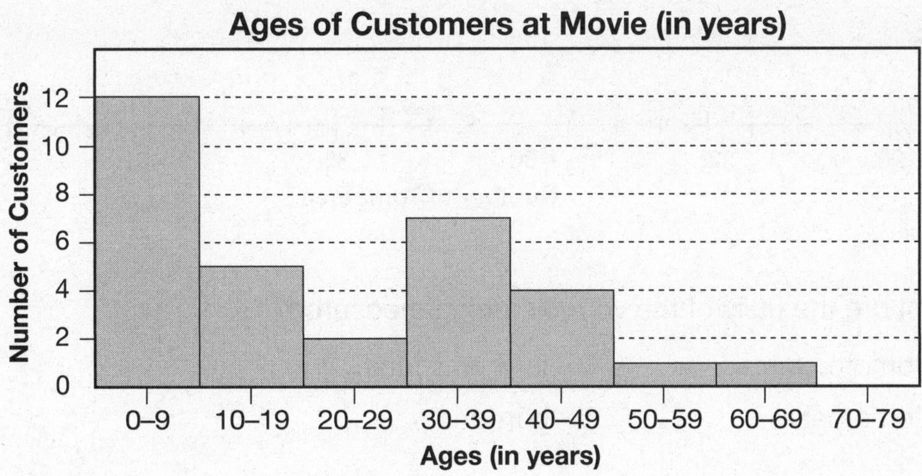

Step 6: Analyze the data.

Because the majority of the data appears on the left side, the histogram is right-skewed. The tallest bar is for 0−9, which means that more customers are between 0 and 9 than in any other age group. This is to be expected for a kids' movie. The shape of the histogram does not show that the data go down exactly with increasing age, however. After the initial peak at 0−9, there is another peak at 30−39, which probably represents the parents of the children.

Practice

Directions: Answer questions 1 through 5 based on the gardener's data set.

A gardener wants to know how many tomatoes a tomato plant yields in one season. The gardener counts the number of tomatoes from 13 plants in a garden. Over the course of the summer, the plants yield the following numbers of tomatoes:

36, 24, 33, 38, 40, 28, 30, 34, 36, 20, 44, 36, 30

1. What are the measures of center of the gardener's data set?

   mean: ____________ median: ____________ mode: ____________

2. What are the measures of variability of the gardener's data set?

   range: ____________ interquartile range: ____________

   mean absolute deviation: ____________

3. Create a box plot based on the gardener's data set.

Number of Tomatoes per Plant

4. What are the units of the variable being measured?

   A. tomato plants
   B. tomatoes
   C. tomato sauce
   D. gardeners

5. Describe the general pattern of the data.

   ____

   ____

CCSS: 6.SP.2, 6.SP.5.a, 6.SP.5.b, 6.SP.5.d

Directions: Answer questions 6 through 10 based on the meteorologist's data set.

A meteorologist in a town in Florida wants to know how many days it rains in the town each month. The meteorologist counts the number of rainy days each month for a year.

Number of Rainy Days per Month							
January	6	April	6	July	17	October	9
February	8	May	9	August	17	November	6
March	8	June	14	September	14	December	6

6. What are the measures of center of the meteorologist's data set?

   mean: ____________ median: ____________ mode: ____________

7. What are the measures of variability of the meteorologist's data set?

   range: ____________ interquartile range: ____________

   mean absolute deviation: ____________

8. Create a dot plot based on the meteorologist's data set.

Number of Rainy Days per Month

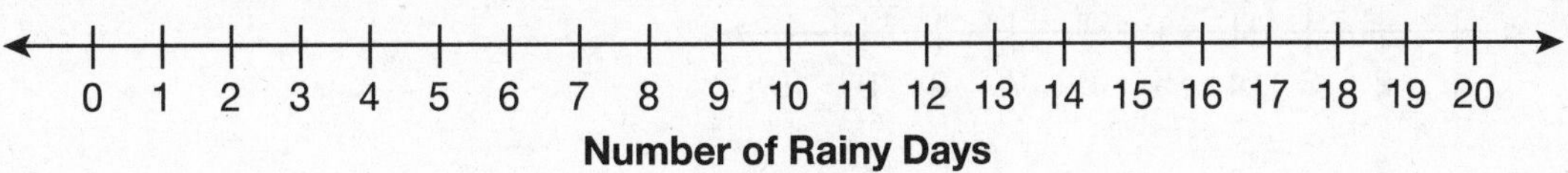

9. What can you conclude from the data set?

   ____

   ____

10. How would you answer the meteorologist's statistical question? Explain your answer.

   ____

   ____

Unit 5 Practice Test

Read each question. Choose the correct answer.

1. Which of the following is an example of a statistical question?

   A. What are the weights of the elephants at the zoo?

   B. How many elephants are at the zoo?

   C. What is the name of the most famous elephant?

   D. How much does the heaviest elephant weigh?

2. A rock band sells the following number of posters at its 12 concerts.

   55, 63, 49, 55, 71, 64, 69, 51, 54, 89, 55, 69

   What is the mean number of posters sold during the rock band's concerts?

   A. 55

   B. 59

   C. 62

   D. 63

3. Which of the following is a measure of variability?

   A. mode

   B. interquartile range

   C. median

   D. average (mean)

4. The nine players in a baseball lineup have the following numbers of home runs.

   8, 2, 7, 2, 9, 1, 7, 5, 2

   Which dot plot represents the data set?

A.

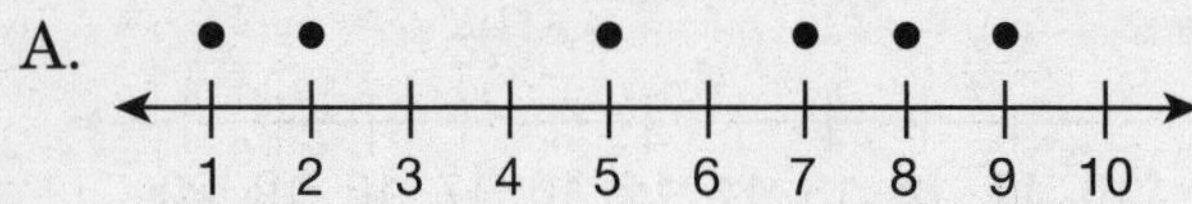

B.

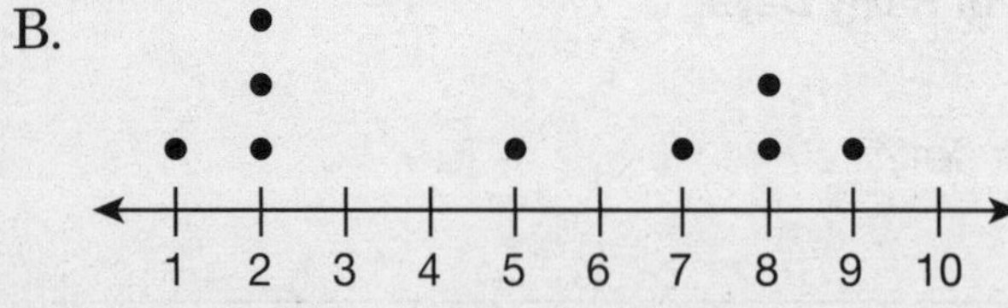

C.

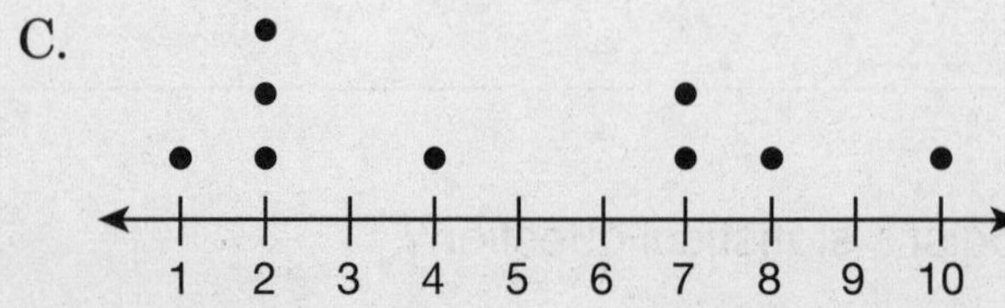

D.

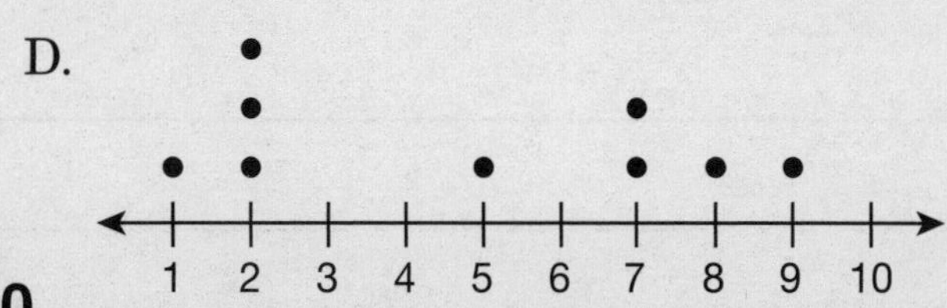

5. The heights, in feet, of 8 houses on a street are listed below.

26, 20, 30, 28, 20, 34, 20, 31

What is the median height, in feet, of the houses on the street? ______________

6. Jermaine wants to find the average distance that each of his friends travels to get to school. What unit of measurement can Jermaine use with his data?

____

7. Abby has 4 cats. Each cat is a different age. Marley is 5, Jibboo is 12, Ernie is 3, and Cassidy is 16. Plot the ages of Abby's cats on the following number line.

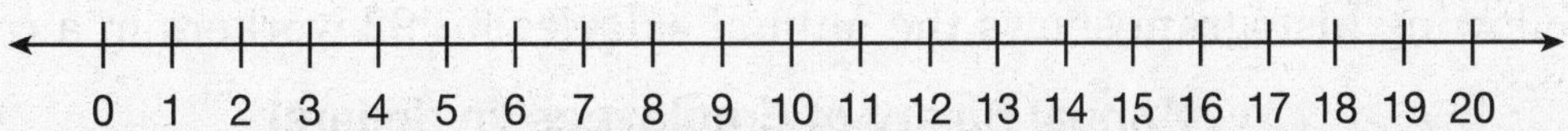

8. The following box plot represents a scientist's data set.

What is the median of the scientist's data set? ______________

9. Is the following question a statistical question? Write *yes* if it is and *no* if it is not.

   What is the minimum age to vote in the United States? ______________

10. The number of eggs laid by 8 different hens in a month is shown in the following list.

    19, 23, 27, 24, 19, 22, 26, 24

    What is the mean absolute deviation of the data set? ______________

11. Does the interquartile range describe a measure of center or a measure of variation?

    ____

12. The following histogram shows the annual salaries for 92 workers in a company.

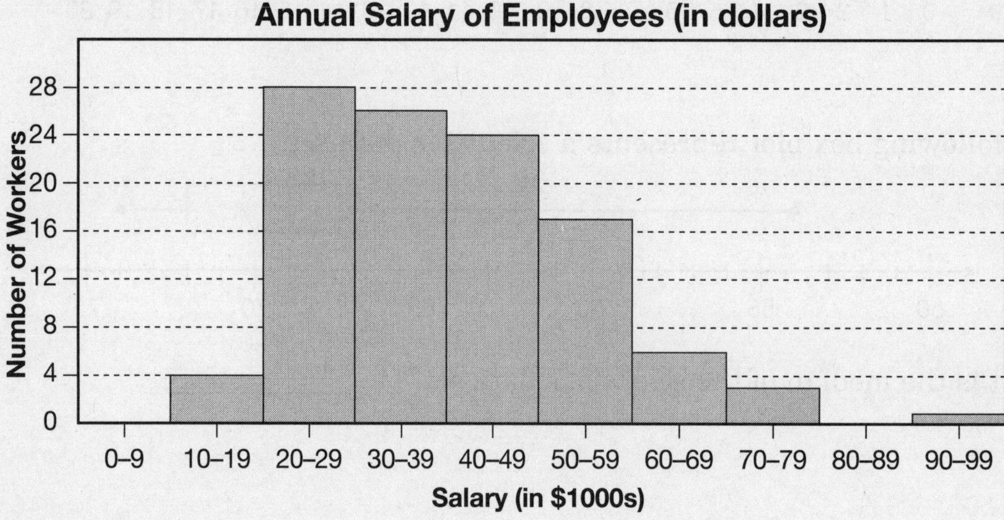

Describe the distribution and characteristics of the data.

____

____

____

13. There are 10 fish tanks at an aquarium. The number of fish in each tank is given in the list below.

32, 38, 29, 40, 52, 27, 35, 44, 41, 31

What is the mode number of fish in the aquarium's tanks? If there is no mode, write *no mode*.

____

14. John measured the low temperature each day during a 15-day period. The following list shows John's temperatures, in degrees Fahrenheit.

58, 49, 41, 56, 44, 45, 46, 60, 46, 46, 55, 52, 53, 59, 53

Create a box plot to show John's data.

Low Temperature (in °F)

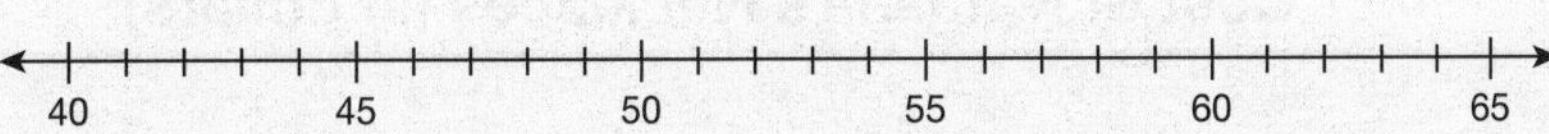

15. Sasha is a student in college. She asks the 20 students in her class for their ages. Then she created the following dot plot.

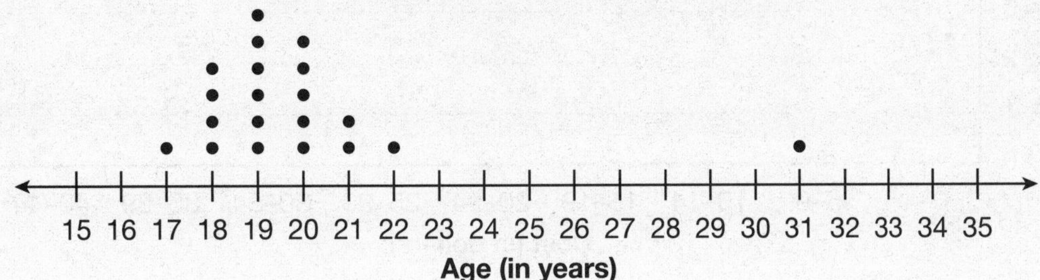

Are there any striking deviations in Sasha's pattern? If so, describe what value or values deviate from the pattern.

____

____

____

16. Is the following question a statistical question? Write *yes* if it is and *no* if it is not.

    What is the cost of a house in the Los Angeles metropolitan area? ______________

17. Alan caught 7 fish on a fishing trip. The weights of his fish are listed below, in kilograms.

    4.3, 3.2, 2.8, 3.1, 2.2, 2.5, 4.3

    What is the mean weight of Alan's fish? ______________

18. Natasha sells homemade necklaces at an arts and crafts fair. She sells 25 necklaces for the following amounts, in dollars.

    22, 5, 10, 8, 24, 12, 19, 29, 28, 24, 12, 15, 16, 40, 18, 34, 20, 22, 27, 28, 30, 32, 35, 38, 20

    Create a histogram to show the Natasha's data set.

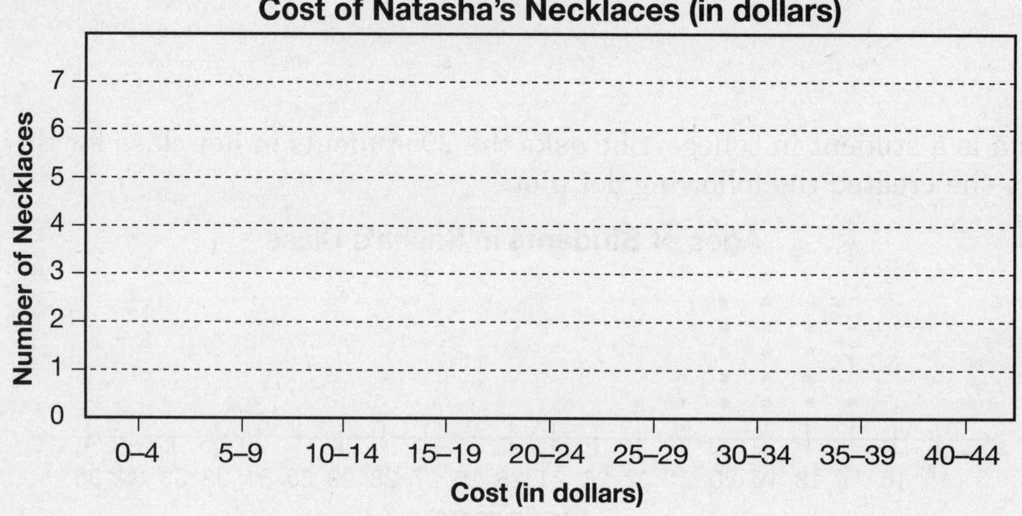

19. The Sanchez family receives an electric bill each month. The amounts of the bills over the last 10 months are listed below, in dollars.

88, 94, 99, 87, 107, 99, 112, 105, 93, 116

What is the mean absolute deviation of the Sanchez family's electric bills?

____

20. A supervisor measures how long 5 customers wait on the phone to speak to an operator. The times are listed below, in minutes.

4, 2, 7, 9, 1

Plot the supervisor's data set on the following number line.

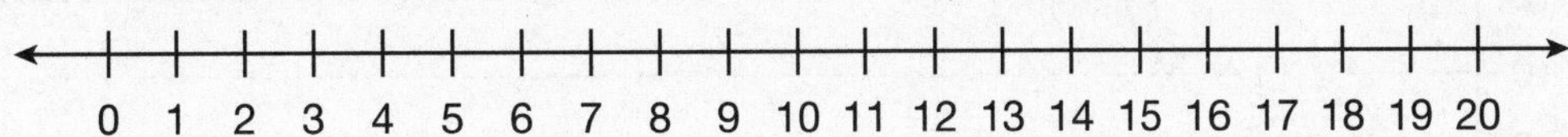

21. Does the median describe a measure of center or a measure of variation?

____

22. Is the following question a statistical question? Write *yes* if it is and *no* if it is not.

What are the weights of the horses on a farm? ______________

23. What is the median of the following data set?

102, 7, 122, 87, 94, 44, 204, 88, 81, 90 ______________

24. A researcher wants to determine the most common age of a person in a town. The researcher asks for the ages of people walking on the street in front of a high school. The following table shows the data.

Ages of People on Street (in years)									
17	28	15	16	42	70	14	18	53	15
46	16	34	17	29	15	16	39	15	14

Part A

Create a histogram based on the researcher's data.

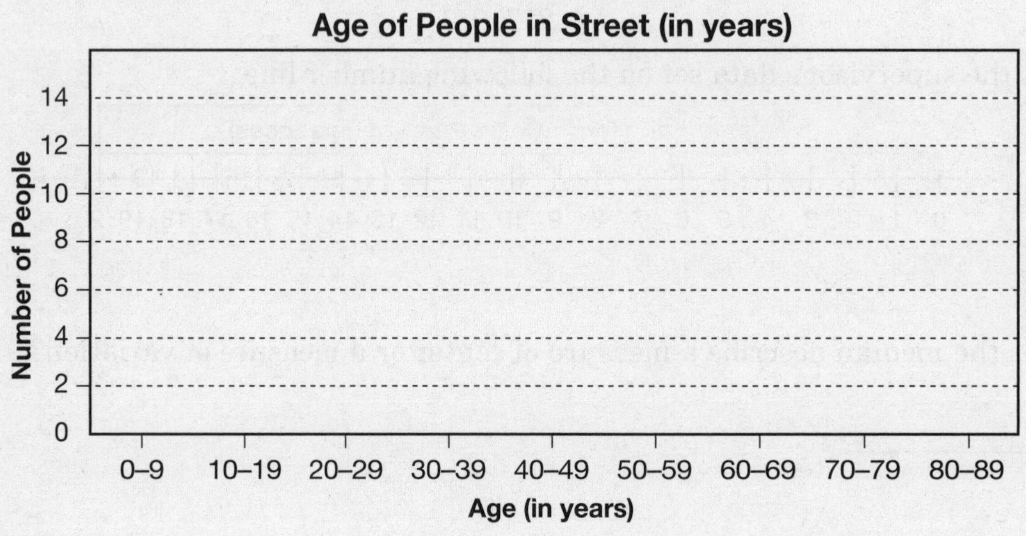

Part B

Describe the distribution of data in the histogram. Use a measure of center, a measure of variability, or a description of the overall shape in your answer.

____

____

____

Part C

How did the way that the researcher got the data affect the results? Explain your answer. Use a measure of center or variability in your answer.

____

____

____

25. A veterinarian wants to answer the following question.

What are the weights of newborn collie puppies?

Part A

Is the veterinarian's question a statistical question? Explain your answer.

____

____

____

Part B

The veterinarian records the following weights of newborn collie puppies during a month.

Weights of Newborn Collie Puppies (in ounces)								
6	9	16	13	14	12	10	16	13
18	14	10	8	5	17	12	14	15

How many observations did the veterinarian make?

____

Part C

Create a dot plot to represent the veterinarian's data.

Weights of Newborn Collie Puppies

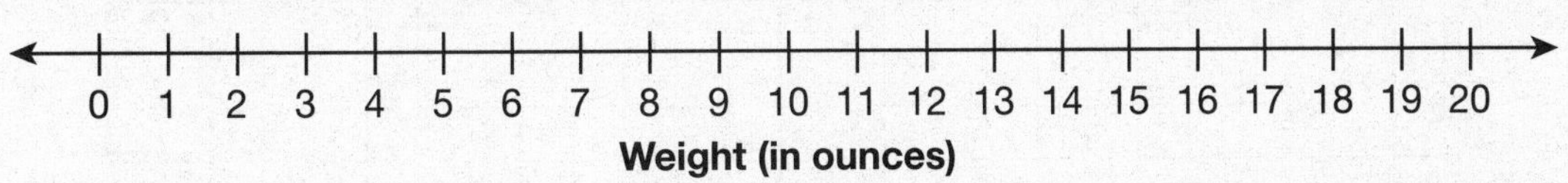

Math Tool: Coordinate Grid

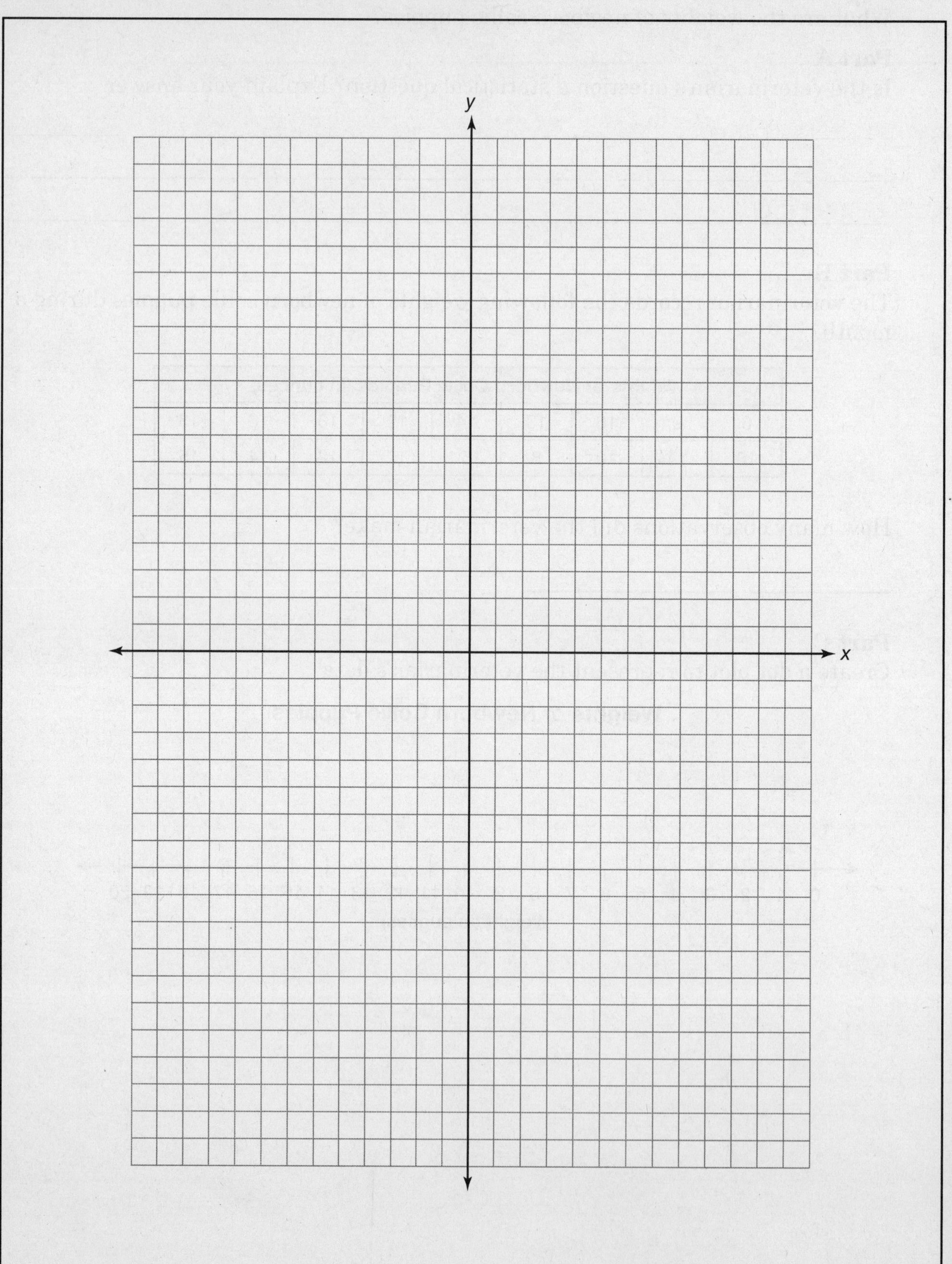

Cut or tear carefully along this line.